T0249528

SOLAR ENERGY

Photovoltaics
and
Domestic Hot Water

SOLAR ENERGY

Photovoltaics
and
Domestic Hot Water

A Technical and Economic Guide for Project Planners, Builders, and Property Owners

Russell H. Plante

AMSTERDAM • BOSTON • HEIDELBERG • LONDON
NEW YORK • OXFORD • PARIS • SAN DIEGO
SAN FRANCISCO • SINGAPORE • SYDNEY • TOKYO
Academic Press is an imprint of Elsevier

Academic Press is an imprint of Elsevier
525 B Street, Suite 1900, San Diego, CA 92101-4495, USA
225 Wyman Street, Waltham, MA 02451, USA

First edition 2014

Copyright © 2014 Elsevier Inc. All rights reserved.

No part of this publication may be reproduced, stored in a retrieval system or
transmitted in any form or by any means electronic, mechanical, photocopying,
recording or otherwise without the prior written permission of the publisher.

Permissions may be sought directly from Elsevier's Science & Technology Rights
Department in Oxford, UK: phone (+44) (0) 1865 843830; fax (+44) (0) 1865 853333;
email: permissions@elsevier.com. Alternatively you can submit your request online
by visiting the Elsevier web site at http://elsevier.com/locate/permissions, and selecting
Obtaining permission to use Elsevier material.

Notice
No responsibility is assumed by the publisher for any injury and/or damage to persons
or property as a matter of products liability, negligence or otherwise, or from any use
or operation of any methods, products, instructions or ideas contained in the material
herein. Because of rapid advances in the medical sciences, in particular, independent
verification of diagnoses and drug dosages should be made.

Library of Congress Cataloging-in-Publication Data
Plante, Russell H. (Russell Howard), 1947–
    Solar energy, photovoltaics, and domestic hot water : a technical and economic guide
    for project planners, builders, and property owners / Russell H. Plante.
      pages cm
Includes bibliographical references and index.
    ISBN 978-0-12-420155-2 (paperback)
1. Solar houses. 2. Solar water heaters. 3. Building-integrated photovoltaic systems. I. Title.
    TH7414.P53 2014
    690'.83704724–dc23
                        2014003358

British Library Cataloguing in Publication Data
A catalogue record for this book is available from the British Library

For information on all Academic Press publications
visit our web site at store.elsevier.com

ISBN: 978-0-12-420155-2

Working together
to grow libraries in
developing countries

ELSEVIER   Book Aid
            International

www.elsevier.com • www.bookaid.org

Dedication

For Kathy—the constant sunshine in my life…

Contents

Preface

There are many books on the market today that address the types of solar energy systems available as well as the methods used to install such systems. Most of those publications, however, do not address the fundamental question of whether or not you should even consider purchasing a solar energy system. This book will help you to understand not only whether or not such an investment will function at your specific location and satisfy your specific demand requirements, but most importantly, to determine if the system *will save you money*. It should not be just a "Green Energy" environmental issue but a decision that encompasses both practicality *and* economics. The information presented throughout these chapters should be of value to a diverse audience not limited to but including facility engineers and managers, project planners, real estate brokers, home builders, alternative energy dealers/installers, educators, and residential and commercial property owners.

The chapters within this book sequentially and independently address topics that will assist you in making an informed decision on whether or not to install a solar energy system for your everyday requirements. This book serves as a prerequisite to the "Do-It-Yourself" installation books that provide guidelines and instructions for installing Solar Domestic Hot Water (DHW) and Photovoltaic (PV) systems. The intention is to provide you with a fundamental understanding of these two types of solar alternative systems in order to make an informed decision as to whether or not such systems are practical for your specific energy demands and economical for you as a long-term investment. This book also provides you information not systematically and comprehensively available on the Internet.

You may already have a basic and fundamental understanding of Solar DHW and/or PV systems, of the components and operations involved, and maybe even a conceptual knowledge of energy costs and comparisons. Perhaps you are only interested in discerning a financial opinion of your investment as to cost payback. In part, this book provides independent topics in a logical progression regarding the evaluation and use of these technologies. In whole, this book will provide you with a fundamental understanding of these systems and guidance to evaluate the basic economic concepts involved, resulting in a comprehensive perspective of these two important alternative solar energy systems.

The predecessor to this book was published by John Wiley & Sons, Inc. in the early 1980s, entitled, "Solar Domestic Hot Water: A Practical Guide to Installation and Understanding". It was written as a practical guide for the homeowner as well as a textbook for the educational market and explained the functions of all the components as well as provided installation guidelines for solar hot water systems. Prior to publication, it received positive technical feedback by nine different college professionals ensuring accuracy and consistency in content and was published when alternative energy solutions were first introduced to the market. This book significantly updates discussions relative to solar DHW and energy conversions, includes technical guidelines for photovoltaics, and explains economic considerations for both types of systems as long-term investments.

During the early 1980s, solar energy alternatives were considered important due to increasing energy prices. When the price of imported oil declined in years following, so did the importance of using solar energy as a cost savings alternative. Federal and most State energy tax credits were eliminated or simply abandoned, and America went back to the old way of doing business by continuing its dependence on foreign oil. The use of solar energy was more or less dismissed by the majority of people, setting the stage for fewer solar collector manufacturers, fewer installation and support businesses, and increased foreign oil energy dependence which exists to this day, over 25 years later. Since that time, we have done very little to develop and use alternative energies or to strengthen our own energy infrastructure. It is simply incredible to believe that so little has been accomplished in the interim. The importance of energy independence and cost savings has not changed over the years. In fact, continued growth in foreign energy dependence has decreased our economic stability. If we continue to do nothing, it should be of increasing concern to all of us that those conditions will become even worse in years to follow.

"Green Energy" has become a lexicon to those people concerned about our environment and our foreign energy dependence as well as to those people who believe the term is overused and extreme in its actual conception. The term "solar" has also received disparaging media coverage because of bad business practices, corporate bankruptcies, manufacturing company failures, extreme government regulations, and the federal government's overexuberance to channel taxpayer funds into poorly chosen "business" decisions. The installation of renewable active energy systems using flat plate or vacuum tube panels for solar DHW systems and the use of photovoltaic systems for the generation of electricity can be viable applications of "green energy". It is the actual cost of such systems, the resultant economic savings, and an understanding of their proper applications that should be fully vetted before you make a decision to conclude whether or not solar is a viable alternative for *your* particular situation.

This book will provide you with a unique compilation of information and explanations not available in other publications. It will help you to determine and understand the basic siting requirements, the amount of energy you need, the amount of solar energy available, the methods of comparing collectors, the number of collectors necessary for either hot water or electricity, and the resulting investment payback for each type of system. Examples of present and future worth of money are also discussed relative to break-even costs and cash flow analysis, providing additional insight regarding these alternative energy system investments. You will then be able to make an informed decision as to the economic practicality of these "Green Energy" sources for your particular project, facility, or home.

Russell H. Plante
Engineering Physicist

Acknowledgments

I would like to thank several people who were helpful during the initial research and editorial phases of this book. I wish to thank Phil Coupe, cofounder of ReVision Energy for the discussions, photographs, and knowledgeable insight he provided relative to both solar domestic hot water (DHW) and photovoltaic (PV) alternative energy systems. I want to thank Christopher Baker, who spent many hours providing very valuable editorial comments throughout the first draft of the original manuscript and Carl Corsello who identified areas of the text needing additional clarifications. I wish to thank Bruce A. Barker, President of Dream Home Consultants, LLC. and author of several books for information, photographs, and changes he recommended and provided. His thoughtful insight was greatly appreciated. Appreciation is also extended to David C. Blakeslee for providing helpful suggestions for technical clarifications prior to publication and to Kattie Washington at Elsevier for her continued editorial support.

I also want to thank Tiffany Gasbarrini, the senior acquisitions editor at Elsevier, for believing in what I had written and recognizing the importance of the book's approach in providing the reader with information necessary to understand the technical and economic value of using solar energy alternatives. She brightened my day when she sent an e-mail message stating, "I think what you've done is unique, and valuable, in emphasizing the economics of solar in a pragmatic, down-to-earth fashion."

And finally, I want to thank my wife, Kathy, for her patience and helpful comments and suggestions throughout the various drafts of this book.

Acknowledgements

Chapter | one

Considering the Solar Alternative

The direct use of solar energy has never been adopted on a worldwide scale because cheaper alternative energy sources always have been available. It really was not that these sources were cheaper than the sun's energy, which is free, but rather that the costs of constructing devices or systems to use the sun's energy were greater than the costs needed to use the other sources that were available. The cost of conventional energy sources will continue to increase, and the reliability of foreign energy imports will continue to be questionable at best. We are overdue in making a serious effort to apply the sun's energy to complement a larger portion of our ever-increasing energy needs. The most appropriate and cost-effective large-scale application of solar energy involves the heating of water for domestic use and the generation of electricity for grid-tied residential use.

Discussions regarding solar domestic hot water (DHW) and photovoltaic (PV) systems have been separated into individual chapters and topic sections in this book so that you can pick and choose the subjects that hold your particular interest. For instance, if you already have the basic knowledge to site and orient a solar DHW or PV system, then simply progress to the next chapter. The information presented in this book is considered to be a prerequisite to purchasing a system and to either having it installed or deciding that additional knowledge is required to install it yourself. The following chapters and their specific sections will sequentially and logically provide you with information so that you can

1. determine whether or not you have the proper site for collecting sufficient solar radiant energy (Chapter 2, Section 2.4, Siting a Solar Energy System),
2. determine what your energy requirements are for hot water (Chapter 3, Section 3.5.1, Hot Water Requirements) and electricity (Chapter 3, Section 3.5.2, Electricity Requirements),
3. determine the amount of solar energy available (Chapter 4, Section 4.5, Determining Solar Energy Availability),
4. determine how to size a solar DHW system (Chapter 4, Section 4.6, Sizing a Solar DHW System) or how to size a solar PV system (Chapter 5, Section 5.4, Sizing a Solar Photovoltaic System),
5. understand the economic criteria involved with making a financial decision (Chapter 6; Economic Criteria for Financial Decisions),

6. evaluate and determine economic payback for hot water (Chapters 7 and 9) and electricity (Chapters 8 and 9), and
7. consider the importance of using renewable energies while understanding implications relative to energy policies and regulations and their effects on the economy (Chapter 10, The Energy Conundrum and Economic Consequences).

Many sections throughout this book provide additional details regarding these two distinct energy alternatives. Worksheets are included to provide a manual (non-Internet) approach to calculating and estimating sizing, energy output, and economic analysis, as well as information relative to Internet online calculators and estimator programs. If you decide that you meet the site conditions necessary to support these solar applications, then other sections of this book should be of further interest by assisting you in making an informed decision regarding purchase and installation.

The cost of energy changes from day to day because of price variations in demand, fuel costs, and availability of generation sources. Whether you heat water by oil, electricity, or other means, or are concerned with the continued price volatility of your electricity demands, a properly installed solar DHW or PV system can save you money. These cost savings are discussed in more detail in Chapters 7, 8, and 9. Simply stated, the first most practical application of "green energy" is the use of solar DHW systems because of the smaller investment expense. Being initially more expensive than solar DHW systems and normally having a slightly longer payback period, PV systems are the second most practical application of solar energy.

The average residential price of electricity in the United States in 2012–2013 was approximately $0.12 per kilowatt-hour (kWh), and in the Northeast, as high as $0.17 per kWh, including line delivery fees. (At times the delivery fees are a large percentage of the total price per kilowatt-hour.) The average price of oil during that same period was $3.69 per gallon. These prices vary over time and by locality because of the availability of power plants, fuel costs, and regulations.

This book's intent is not to elaborate on conservation measures or the assessment of the variety of alternative energy applications, but rather to develop an understanding of the practical and economic applications of solar DHW and solar PV systems. In the past, homeowners have not had adequate assurance or knowledge about either solar DHW systems or PV systems, or their operation and proper installation. Today, such information is available in abundance, almost to the point at which it can be confusing.

Many people are quite capable of installing their own systems, thereby defraying installation costs. Such installations, however, require the do-it-yourselfer to be multidisciplined in carpentry, plumbing, electrical, and solar, meeting all pertinent local and state codes. In particular, PV installations require an in depth understanding of Article 609 of the National Electrical Code to ensure electrical safety requirements are met. In addition, some states that offer tax incentives require that a solar energy system be installed by a certified installer to receive a state tax rebate. Even if a person intends to have the system installed professionally, they still should have an understanding of the various types of systems available to make an intelligent decision. The following chapters, therefore, provide a logical approach to evaluating and understanding energy cost comparisons per British thermal units, to determining proper siting, sizing, and availability of solar insolation, and to understanding some of the basic component configurations and operating fundamentals.

Time after time the same arguments against the use of solar energy systems have appeared, and these beliefs have slowed their adoption, resulting in the delay of our independence from oil. These arguments include the following:

1. Solar cannot work in many of the northern sections of the United States.
2. Solar energy systems are not a good investment and will never pay for themselves.
3. Equipment will be cheaper in the future.
4. Solar technology is untried and not perfected.
5. Solar energy systems are difficult to install.

All of these statements are false. One should carefully consider the following:

1. Solar energy utilization is feasible in every part of the country. The amount of radiant energy (*insolation*) received in the Northeast is only slightly less than the national average, still providing sufficient solar energy to supply hot water and electricity.
2. Solar DHW is the most cost-effective use of solar energy and can be justified economically. With current energy sources growing increasingly expensive, a solar DHW system will increase the value of the home as it reduces the utility bills. In most cases, payback is within 5–10 years depending on system type and whether or not the system is self-installed, thereby saving labor costs. Solar PV systems normally require a slightly longer payback period, but the percentage savings can be calculated more readily than other solar applications because there are fewer parameters to consider and because the demand for energy consumption is defined more strictly.
3. Solar equipment is becoming *more* costly because of increasing material and labor prices. Furthermore, recurrent federal and some state tax credits make this an opportune time to purchase equipment now, not later.
4. Solar DHW technology is a safe, proven, and reliable low-energy technology that has been used since the early 1900s in this country and earlier in others. The technology for solar DHW is well developed. Standard flat-plate collectors, storage tanks, and control systems are commercially available. In the past, issues such as faulty and careless installation, poor workmanship, and the improper use of components and materials have produced consumer skepticism. Engineering detail and design integrity also have been ignored previously in many instances, resulting in either system inoperability or in poor system performance. These unfortunate situations have become more isolated because of the increased field expertise of the solar installer or

dealer. You simply have to engage a reputable company to ensure that an installation is completed properly, not unlike any other contractor you would hire. The technology is proven.

In addition to solar DHW, solar PV technology has made gigantic strides in the past several years in the conversion of light to energy and the variety of solar cells available. Twenty-first-century PV modules have a 25 plus years of life with little degradation and have been proven to be a reliable alternative energy application. Research in this area continues, including such advancements in studies involving "quantum dots" as well as "thermophotovoltaics" and their impact regarding improvements in PV efficiency. Technology will always continue to evolve.

5. Solar energy systems are not difficult to install. It is, however, necessary that these systems be properly sized, sited, mounted, equipped, and maintained to ensure their economic viability. It is recommended that the homeowner have such systems installed by a reputable certified dealer. For those people who are construction savvy, however, very good installation guide books are available. Although this book contains information on siting and sizing a system to address the economic feasibility of using solar as a "green energy" solution, it does not address detailed installation procedures. Its main premise is to provide a technical and economic set of guidelines.

The use of solar energy to heat water for domestic use as well as to provide electricity is questioned because it is unfamiliar and not considered to be conventional. There appears to be a mental block against using new methods to replace the conventional ones because of the lack of education. Everyone is familiar with the older methods using gas, electric hot water tanks, and furnaces, and feels "comfortable" with them. It is now time, however, to understand the newer methods of employing these alternative energies. Increasing the use of solar DHW and solar PV is a viable first step in energy independence, which also reduces fossil fuel consumption, leaving more fossil fuels in the ground, and thereby lessening our carbon emissions.

Cost considerations can be made by contacting a state certified installer or dealer to provide a design plan and purchase estimate, normally free of charge. Before you request an estimate, however, it is advisable that you understand a few basic solar and investment fundamentals. You can then make a comprehensive decision as to whether or not you can benefit from the use of solar energy. In other words, it is always a good idea to have some knowledge about a subject when you are going to discuss it with someone, simply because ultimately you will have a more in-depth conversation and better understanding before making a financial commitment. If it is not a practical application in both function and cost, then solar is *not* a solution. If it is concluded that solar might be a viable alternative after reading this book, then at least an informed decision can then be made on whether or not to invest in a system.

The sun has provided us with stored chemical energy in the form of fossil fuels, which constantly is being depleted, and this depletion is responsible for escalated social and economic issues. To curtail these adversities, the direct application of the sun's radiant energy to alternative conversion processes such as PV, photochemical, thermoelectric, and heat must be developed and utilized continuously. An economic first application involves the use of solar collectors to convert the sun's radiant energy into heat energy for domestic water heating and electricity. We should now take the opportunity to use the most vast, continuing energy resource available to us on a worldwide basis: our sun.

Chapter two

Simple Solar Basics

2.1 SUN AND EARTH FUNDAMENTALS

The sun is everyone's life. Without its energy, our past, present, and future would not be. In brief retrospect, our primordial star is theorized to have been formed from a cosmic cloud of individual particles in which matter was concentrated under the action of gravitational energy. This large mass of gas condensed, and as the gravitational and kinetic energies of the particles increased, the interior of the cloud heated to extremely high temperatures. Finally, increased temperatures and pressures in the interior of this dense mass reached proportions capable of sustaining the nuclear reaction known as fusion. It is this energy that has sustained human existence.

For nearly 2000 years, virtually everyone thought our earth was the center of the universe and that the sun, and all the other heavenly bodies, revolved about our planet. We now know that the apparent motion of the sun across the sky is actually the result of the earth's own rotation. The earth spins on its axis at a rate of approximately 360.99° in 24 h, and, therefore, the sun appears to move across the sky at a rate of about 15.04°/hr. The earth moves about the sun in an approximately circular path with the sun positioned slightly off center along the long axis of its ellipse. This offset is such that the earth is closest to the sun around January 1, and farthest from the sun around July 1. Our changing seasons occur because the earth's rotational axis is tilted at approximately 23.5° with respect to the plane of the ecliptic containing our orbit. Figure 2.1 represents the earth–sun relationship, as it would be viewed by an observer far out in space.

From Figure 2.1, it can be seen that in the northern hemisphere, the north end of the axis is tilted away from the sun during the winter and that the north end of the axis is tilted toward the sun during the summer. In the southern hemisphere, the seasons are reversed and the tendency is for greater seasonal differences in temperature than experienced in the northern hemisphere.

The angle of the earth's tilt with respect to the sun and the equatorial plane, shown in Figure 2.1, is called the *declination angle*. This angle varies throughout the year from +23.5° on June 21 to −23.5° on December 21. When the earth's axis is perpendicular to the line joining the earth and the sun, day and night are of equal length (March 21 and September 21) and are called the *spring and fall equinoxes*, respectively. When the angle of declination is at its greatest at +23.5°, a point in the northern hemisphere will have its longest period of daylight, called the *summer solstice* (June 21). When the angle of declination is at its lowest at −23.5°, a point in the northern hemisphere will have its longest period of darkness, called the *winter solstice* (December 21).

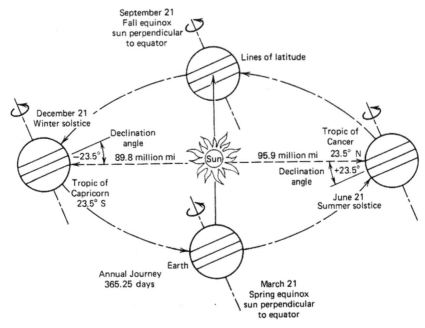

FIGURE 2.1 Earth–sun relationship.

Lines of latitude, which normally are included on most maps, are actually designations of the angle between the equator and a line from the center of the earth to its surface as shown in Figure 2.2. Latitudes vary from 0° at the equator to 90° at the earth's poles and are parallel with the equator. The latitude denoting the most northerly position of the sun when the declination angle is +23.5° is know as the *Tropic of Cancer*. The latitude denoting the most southerly position of the sun when the declination angle is −23.5° is known as the *Tropic of Capricorn*. For the purpose of determining time, an imaginary circle around the earth at zero degrees latitude has been divided into 24 segments of 15° each. These circular lines are called lines of longitude and extend from the North Pole to the South Pole. Lines of longitude are defined to start at zero degrees from Greenwich, England.

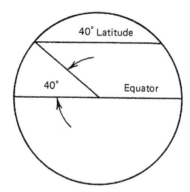

FIGURE 2.2 Lines of latitude.

The apparent position of the sun from any point on earth is defined by two angles. The angle of the sun's position in the sky with respect to the earth's horizontal is known as the solar *altitude*. The position of the sun with respect to true south is referred to as the solar *azimuth*. Figure 2.3 illustrates typical altitude and azimuth positions of the sun at the equinox and solstice days. When the sun's position is *true south*, the azimuth angle is zero and the altitude angle is a maximum at a time referred to as *solar noon*. The term *true south* will be discussed during the explanation of siting a solar energy system.

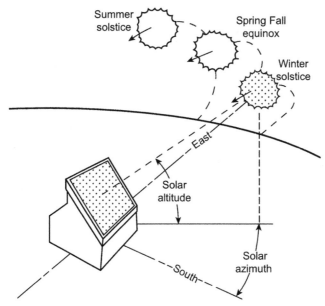

FIGURE 2.3 Typical altitude and azimuth positions. Courtesy of Copper Development Association, Inc. New York, New York.

The position of the sun in relation to specific geographic locations, seasons, and times of day can be determined by several different methods, each with varying degrees of accuracy. Graphic projections, however, can be understood easily and can be correlated to both radiant energy and shading calculations. These projections are referred to as **sun path diagrams** that depict the path of the sun within the "sky vault" as projected onto a horizontal plane. From these sun path diagrams, we can generate what is called a **Mercator projection**, which graphically depicts altitude and azimuth for each month onto a flat map for each variation of latitude. This concept may sound confusing at first, but it will be illustrated and explained in this chapter during the discussion on siting a solar energy system.

2.2 RADIANT ENERGY CONSIDERATIONS

Our sun belongs to a class of dwarf yellow stars of spectral type G. It has a diameter of approximately 864,000 miles, which is greater than three times the distance from the earth to the moon, and it rotates on its axis from west to east with a period of rotation at its equator of approximately 27 days. The sun's mass is approximately 2×10^{30} kg, its volume is approximately 1.4×10^{27} m³, and its density is approximately 1.4 g/cm³. The basic structure of our sun is composed of seven layers. From interior outward, these layers are designated as the core, radiation zone, convective zone, photosphere, reversing layer, chromosphere, and corona. The corona is the final region of the sun and can be seen during a total solar eclipse as a pale white halo extending one or two sun diameters beyond the sun, where the temperature is approximately 4,000,000 °F.

The principal source of the sun's radiant energy is the fusion of hydrogen nuclei, which leads to the formation of helium. Nuclear fusion involves the combining of several small nuclei into one large nucleus with the subsequent release of huge amounts of energy. The nuclei of hydrogen are single particles called protons, each of which carries a positive electric charge. Similarly charged particles normally repel each other, but if the temperature is high enough, their motion can be sufficiently vigorous to allow them to approach very closely, and this short-range attractive force can result is fusion. For our sun, this fusion reaction is known as the proton–proton reaction. The conversion of hydrogen to helium actually involves three separate reactions. In summarizing this process, we can say that four hydrogen nuclei combine to form helium and a large amount of energetic radiation called gamma rays. This outgoing radiation is a very high-energy electromagnetic radiation, which has the shortest wavelength known—approximately a hundred-millionth of a millimeter. As the initial gamma radiation strikes nuclei and electrons, or is scattered in near-collisions, varying forms of energy result, which have less energy than the original gamma radiation but with energy in longer wavelengths, such as X-rays, ultraviolet, visible light, infrared (heat), and radio waves. These radiant energy waves can be arranged in increasing order of wavelength as illustrated in Figure 2.4.

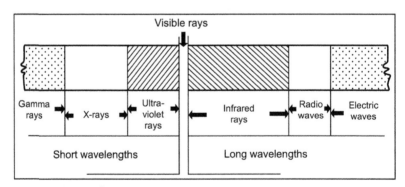

FIGURE 2.4 Electromagnetic spectrum. Department of Energy.

The most general law of radiation in called Plank's law, in which the energy content of electromagnetic radiation is represented by $E = hf$, where E is energy, h is a constant (Plank's), and f is the frequency of the wavelength. The relationship between wavelength λ, velocity v, and frequency f, is represented by the expression, $v = f\lambda$. Because the velocity of the electromagnetic radiation travels at the speed of light, ς, and this speed is equivalent to the product of frequency and wavelength, Plank's law can be rewritten as $E = h\varsigma/\lambda$. The curve shown in Figure 2.5 illustrates the relationship between the amount of energy emitted and the wavelength for objects at a specific temperature. At any one temperature, a wide spectrum of wavelengths is produced. As the temperature increases, so too does the frequency, and this increased frequency results in a shorter wavelength. Therefore, the higher the temperature, the more the maximum energy (peak of the curve) shifts toward the shorter wavelength.

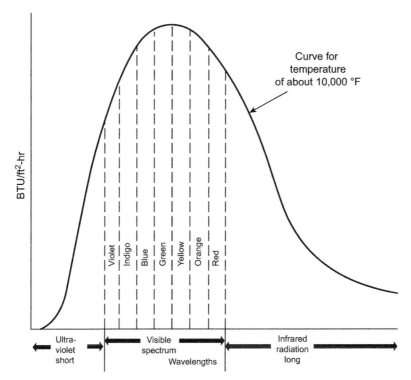

FIGURE 2.5 Radiant energy from the sun.

From this discussion, one can understand that we are concerned with three major energy regions of radiation from the sun. These regions include ultraviolet, visible, and infrared radiation. A solar domestic hot water (DHW) collector absorbs radiation from all three of these regions. In the case of a photovoltaic (PV) module, the spectrum of useful light varies based on the conducting material of the PV cells.

2.3 ENERGY DISTRIBUTION

Before reaching the earth's atmosphere, there is very little loss in the amount of radiation emitted from the sun. Once the sun's energy enters the atmosphere, the various wavelengths of energy are selectively depleted as illustrated in Figure 2.6. Some of the energy is reflected back into space as it enters the upper atmosphere. Energy depletion continues with ultraviolet radiation being absorbed by the upper layer of ozone (O_3), resulting in only 1–3% of the total energy received on the earth's surface being ultraviolet. Only a small amount of absorption occurs in the visible region. Absorption of the longer wavelength radiation (infrared) is due principally to molecules of water and carbon dioxide. Scattering of the shorter wavelengths of visible light causes the sky to appear blue and the sun to appear yellow or orange. This scattering of light is called *diffuse radiation*, and on cloudy days, this may represent all of the solar energy available for use. Most of the visible light, however, that does manage to penetrate the atmosphere is called *direct beam radiation*.

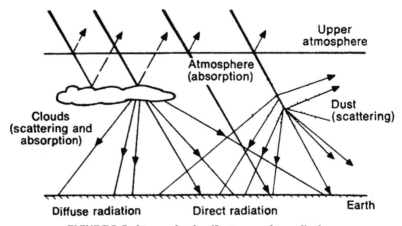

FIGURE 2.6 Atmospheric effects on solar radiation.

The total solar energy received at the earth's surface is called *insolation* (not to be confused with the term *insulation*). Insolation is composed of direct radiation as well as diffuse radiation. The factors that affect the amount of insolation available at a particular location include latitude, time of day, time of year, cloud cover, shading obstructions, atmospheric turbidity, elevation, and orientation of the land surface. The amount of insolation available per unit area of ground surface is determined by the actual depletion of available radiation through the atmosphere and by the angle that the sun's rays make with a surface. When a beam of energy strikes a surface with a cross-sectional area of 1 ft^2 with an angle of incidence ($\angle i$) of 0°, its energy is distributed over an area of 1 ft^2. The angle of incidence ($\angle i$) is depicted in Figure 2.7 as the angle measured between the incoming beam of energy and a line drawn perpendicular to the surface that it strikes.

As the angle of incidence increases to ($\angle i_1$) and ($\angle i_2$), the beam's energy is decreased per unit area as a progressively larger area is covered. This illustrates that the sun's apparent position in the sky is very important for the collection of radiant energy. The more perpendicular or normal the rays of energy are to a surface, the more energy there is per unit area. From the previous discussion of

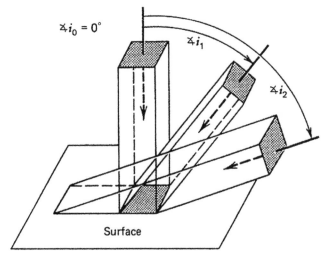

FIGURE 2.7 Energy distribution and angle of incidence.

latitude, it is understood that the higher the latitude, the more slanting are the sun's rays to the surface of the ground, resulting in less energy received per unit area. As the angle of incidence increases, less energy also is received because of the greater depth of atmosphere encountered.

Insolation measured on a horizontal surface plotted against time of day is shown in Figure 2.8. As expected, the total insolation received is maximum at solar noon because the angle of incidence of the sun's rays is minimal.

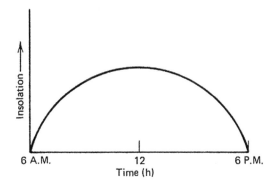

FIGURE 2.8 Insolation versus time.

Because the sun is low in the southern sky during winter in the northern hemisphere, more solar energy will strike a solar collector if it is tilted up from the horizontal at a steep angle toward the sun. During summer, the same collector will intercept more solar radiation in a more horizontal position. The angle of tilt, therefore, is very important in the overall collection of energy. Obviously, the optimum tilt occurs when the angle of the collector is the same as the incoming solar radiation, as illustrated in Figure 2.9.

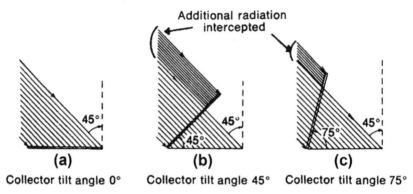

(a) **(b)** **(c)**

Collector tilt angle 0° Collector tilt angle 45° Collector tilt angle 75°

FIGURE 2.9 Effect of collector tilt on energy intercepted.

In general, the complexity involved in making a solar collector adjustable in its tilt angle is not compensated by improved performance. In most situations, it is preferable to select a fixed suitable angle based on the function for which the collector is intended to serve. A tilt angle equivalent to the latitude best serves the needs for the overall year-round collection of energy.

2.4 SITING A SOLAR ENERGY SYSTEM

Before you make the decision to install a solar DHW or PV system, you must determine whether or not you have the proper site so that the system will operate efficiently and productively. Siting a solar energy system is the first step we should perform before discussing energy requirements to determine whether or not a system would be economical. If you do not have the correct geographic location and availability of sunshine, then an economic analysis is moot. Once you determine whether or not you have an appropriate location, you then can determine the amount of energy you need and the size of the collector array needed to produce that energy. Once the size of the system has been determined, the economics of such an investment can be evaluated.

Siting a collector system is comprised of three factors including (1) collector orientation, (2) collector tilt, and (3) collector shading. These siting factors are common to both solar DHW and PV systems.

2.4.1 Collector Orientation

Ideally, the collector array should face the middle of the sun's daily path within ±15° east or west (azimuth) of *true south* as illustrated in Figure 2.10. A quick and fairly accurate method used to determine true south is by using a compass and an *isogonic chart* to compensate for the earth's magnetic field.

True south should not be confused with compass or magnetic south. Because the earth's magnetic field is not aligned parallel with the earth's north–south axis, a compass will not read true. This magnetic declination will vary at each location on the earth's surface. Points on the earth's surface that have the same magnetic declination can be joined together by an imaginary line to form what is called an *isogonic chart*. When the magnetic declination is zero, true and magnetic north

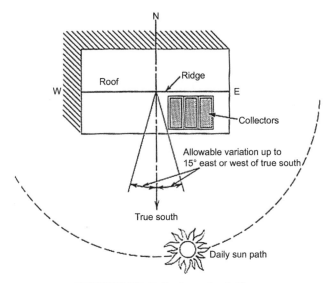

FIGURE 2.10 Collector orientation.

are the same. In the United States, a line of zero declination runs from the eastern end of Lake Michigan through the western edge of Florida to the Gulf of Mexico. On the west side of this zero declination line, your compass needle will point to the east of true north. On the east side of this zero declination line, your compass needle will point to the west of true north. This is illustrated in Figure 2.11.

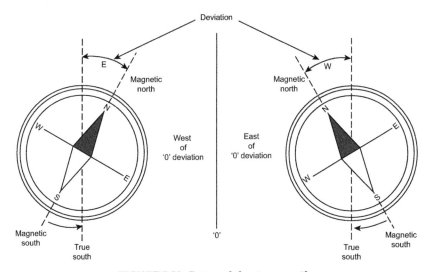

FIGURE 2.11 Determining true south.

The correction factors needed to adjust your compass reading for most locations in the United States can be obtained from the isogonic chart of Figure 2.12. There are slight variations to this type of chart. As an example to correct our compass reading, we find from Figure 2.12 that Billings, Montana, has

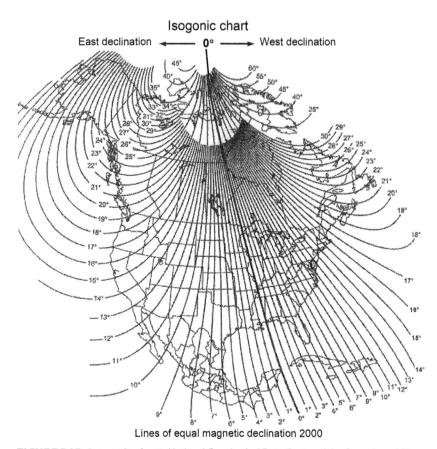

FIGURE 2.12 Isogonic chart. National Geophysical Data Center of the Oceanic and Atmospheric Administration; A worldwide isogonic map is available from the National Geospatial Intelligence Agency at http://msi.nga.mil/MSISiteContent/StaticFiles/Files/mv-world.jpg.

a 15° east deviation from the zero declination line. This means that another 15° east of magnetic south should be added to the compass reading. On the other hand, a family in Boston, Massachusetts, has a 14° west deviation. This means that 14° west of magnetic south should be added to the compass reading. When taking these compass readings, one must remember not to stand near large metallic objects or power lines so as not to distort the earth's lines of magnetic flux and therefore affect the compass reading. More refined correction factors can be obtained by referencing information from the National Geophysical Data Center of the Oceanic and Atmospheric Administration by referring to the Internet online calculators for "magnetic fields" at www.ngdc. noaa.gov/geomag.

To reiterate, we should require an orientation of the collectors to within ±15° east or west of true south. Orientations outside of these parameters can be used but should be evaluated carefully based on available surface area, shading issues, and tilt angle. For solar DHW systems, orienting the collectors toward the east will start the system earlier in the morning, but orientation slightly to the west will increase system performance because ambient temperatures usually are higher in the afternoon, with the collectors consequently losing less heat to the surroundings. If the orientation requirement cannot be met, then additional collectors should be considered, increasing the size of the system compensating the difference in the amount of energy received. For PV systems, it is best to orient the collectors closer to true south, avoiding higher ambient temperatures.

2.4.2 Collector Tilt

The collector array should face the middle of the sun's seasonal path, which is an angle from horizontal equal to the latitude. The rule generally followed for the tilt of a collector array in the northern hemisphere is to position the collector at the angle of latitude, ±10°, as shown in Figure 2.13. Variations of 10° in either way will not seriously affect the total annual collection of the system. Winter system performance, however, can be optimized with the collectors at a tilt of latitude plus 10°, because less energy collection time is available during this period and also because heat loss is greater as a result of lower ambient temperatures.

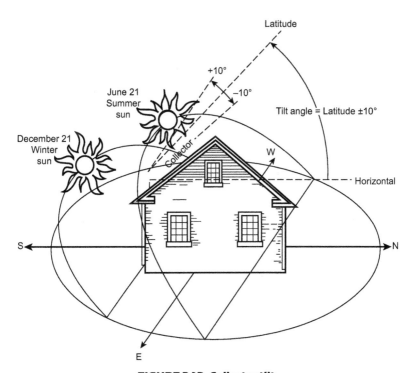

FIGURE 2.13 Collector tilt.

2.4.3 Collector Shading

Shading is a very important factor. No more than 5% of the collector array should be shaded between 9 A.M. and 3 P.M., when the greatest solar potential exists. By knowing the altitude and azimuth of the sun throughout the year, you can determine if a shading problem might exist for your particular site. *Daily* and *seasonal* variation in the angle of the sun as illustrated in collector *orientation* and collector *tilt*, respectively, are summarized in the previously mentioned concept called the sun path.

A "solar window" is actually a plot of the sun's path during the year at a particular latitude. We can derive what is called a *Mercator projection* from these sun path diagrams, which graphically depicts altitude and azimuth for each month onto a flat map for each variation of latitude. Figures 2.14(a)–(i) are Mercator projections that have been replotted from sun path diagrams. These maps are useful for evaluating the site for possible shading restrictions with respect to the solar window. Shading issues can have considerably more impact on the energy output from PV modules than from solar DHW collectors depending on

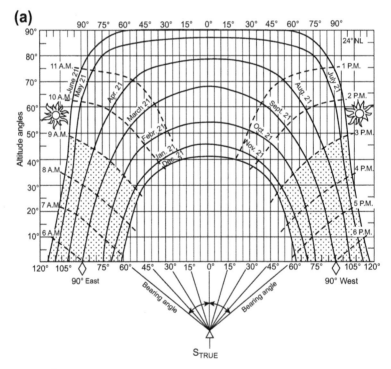

FIGURE 2.14 Mercator projection for latitudes from 24° to 56° north latitude (NL) at 4° intervals. (a) 24° north latitude.

figure continued on next page

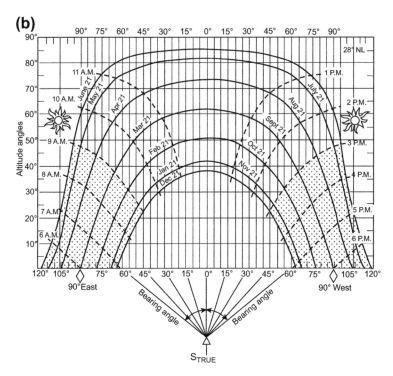

FIGURE 2.14 (b) 28° north latitude.

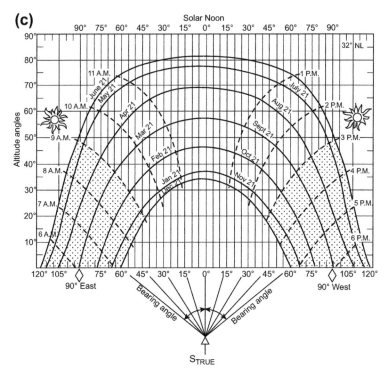

FIGURE 2.14 (c) 32° north latitude.

figure continued on next page

(d)

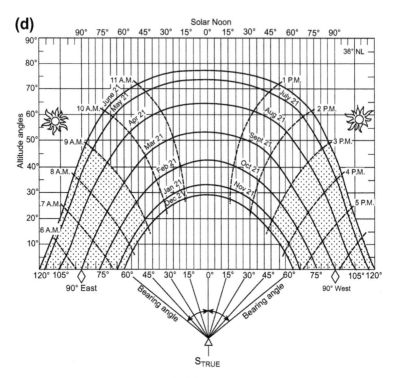

FIGURE 2.14 (d) 36° north latitude.

(e)

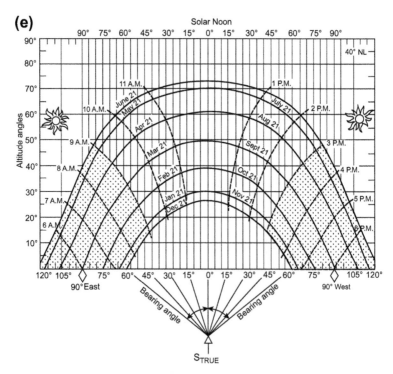

FIGURE 2.14 (e) 40° north latitude.

figure continued on next page

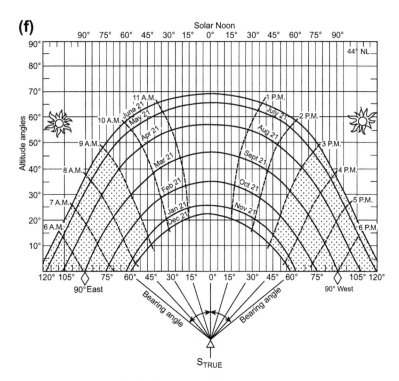

FIGURE 2.14 (f) 44° north latitude.

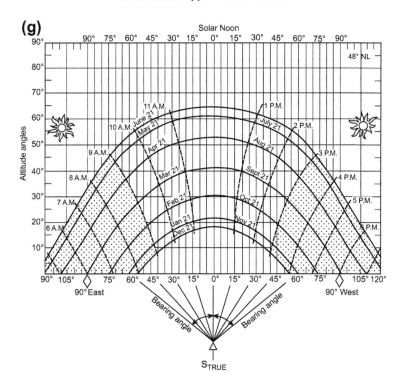

FIGURE 2.14 (g) 48° north latitude.

figure continued on next page

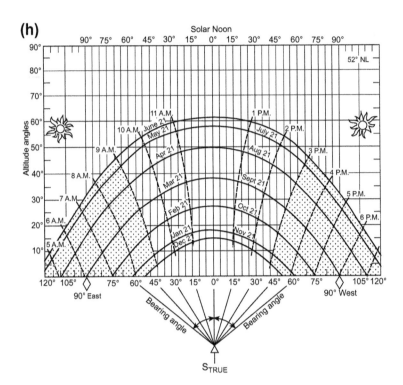

FIGURE 2.14 (h) 52° north latitude.

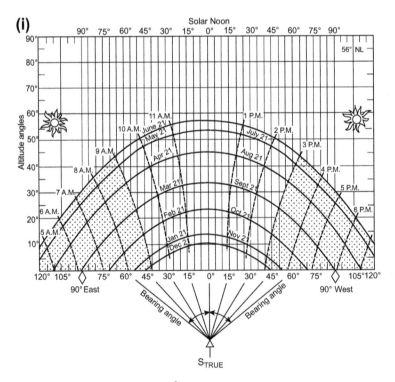

FIGURE 2.14 (i) 56° north latitude.

the type of voltage inverters used in the PV system. Chapter 5, Section 5.2.3, elaborates on the importance of this issue.

A useful instrument in siting solar energy systems is the Solar Pathfinder™, shown in Figure 2.15, that can accurately assess shading patterns and incorporates both the appropriate Mercator projections as previously discussed and magnetic declination adjustments necessary to make an accurate assessment.

FIGURE 2.15 Solar Pathfinder. Photo courtesy of The Solar Pathfinder Company. (For color version of this figure, the reader is referred to the online version of this book.)

Another instrument that can be used to determine the altitude of potential obstructions is the abney level as shown in Figure 2.16. This is a versatile instrument because it can be used to measure slope, to determine height of objects

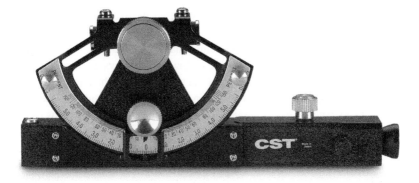

FIGURE 2.16 CST/Berger topographical abney level. Photo courtesy of Forestry Suppliers, Inc. used by permission No. 580082.

(i.e., trees, poles), to determine elevation in relation to a point of known elevation, and to run lines of levels.

By standing at your potential collector site and checking the altitude angle (in degrees) of all objects with an abney level, you can plot the results directly onto a Mercator projection for your latitude as shown in Figure 2.17.

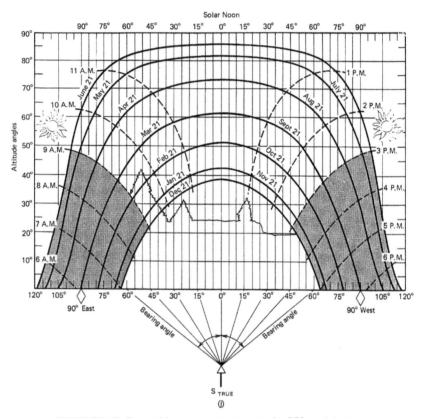

FIGURE 2.17 Typical Mercator projection plot for 28° north latitude.

One of the major sources of shading is caused by trees, so the homeowner should consider the effect of future growth. Chimneys, dormers, adjacent roof sections, fences, topography, and other buildings may shade the collector array, especially in the winter when the sun's angles are low and shadows are long. One must be careful when observing the full path of the sun to prevent shading. By knowing the altitude angle and azimuth of the sun throughout the year and with the use of Figures 2.14(a)–(i), you can accurately assess possible shading problems. At this point, you now will know whether or not you have the proper location and orientation for a solar collector system array. If you have a proper site, then you can proceed to the next step and determine your energy requirements as discussed in Section 3.5 of Chapter 3. Once you know what those requirements are, you should then determine the amount of ***insolation*** available in your area or sun hours in a day to size your system as discussed in Chapters 4 (Sections 4.5 and 4.6) and 5 (Section 5.4). Once you determine the number of collectors that are required to meet your energy requirements for either hot water or electricity, you will be able to calculate the investment cost of the system.

Chapter three

Determining Energy Usage

3.1 ENERGY COSTS AND COMPARISONS

A direct comparison of conventional fuel costs (i.e., gas, oil, electricity) versus solar energy costs can be established by determining the *cost of fuel per unit energy*. Oil, gas, electricity, coal, and wood have different terms of quantity (i.e., gallons, cubic feet, kilowatt-hours, tons, and cords, respectively) that must be related to one common denominator: *cost per British thermal unit* (BTU). A *BTU* is enough heat to heat 1 lb of water 1 °F. The cost of each type of fuel per million BTUs (MBTUs) can be compared with one another using the examples in Table 3.1.

Let's take a look at calculating energy equivalencies. For example, if a source of mixed hardwood is available at $235/cord, and fuel oil is available at

Table 3.1 Fuel Cost Comparisons–Fuel Conversion Assumptions

Type of Fuel	Unit of Measure	BTU Equivalency (Approximate)	Average Assumed Efficiency Use
Fuel oil	1. Gallon 2. 42 US gallons/barrel	138,800 BTUs/gallon	75–83% with heat loss considerations
Liquid petroleum gas	Gallon	91,300 BTUs/gallon	79%
Electricity	Kilowatt-hour (kWh)	3414 BTUs/kWh	98%
Mixed hardwoods	Cord	20 million BTUs/cord	50–79%
Wood pellets	Pound	8200 BTUs/lb (dry pellets)	83–85%
Natural gas	1. Cubic foot 2. Therm	1. 1028 BTUs/cubic foot 2. 100,000 BTUs/therm (29.3 kWh)	75%
Mixed softwoods	Cord	15 million BTUs/cord	50–78%

BTU, British thermal unit.
Source: Data developed from USDA Forest Products Laboratory

$3.50/gallon, a cost per unit energy comparison can be derived using the following energy assumptions:

Mixed Hardwood

$$\frac{20 \text{ MBTU}}{\text{cord}} \times 0.5 \text{ (efficiency)} = 10 \text{ MBTU/cord}$$

$$\frac{\$235}{\text{cord}} \times \frac{1 \text{ cord}}{10 \text{ MBTU}} = \mathbf{\$23.50/MBTU}$$

Fuel Oil

$$\frac{138{,}800 \text{ BTUs}}{\text{gallon}} \times 0.75 \text{ (efficiency)} = 104{,}100 \text{ BTUs/gallon}$$

$$\frac{104{,}100 \text{ BTUs}}{\text{gallon}} \times \frac{1 \text{ MBTU}}{1{,}000{,}000 \text{ BTUs}} = 0.104 \text{ MBTUs/gallon}$$

$$\frac{\$3.50}{\text{gallon}} \times \frac{1 \text{ gallon}}{0.104 \text{ MBTUs}} = \mathbf{\$34.65/MBTU}$$

Therefore, the equivalent energy cost of the mixed hardwood in comparison with the fuel oil is $23.50/MBTU versus $34.65/MBTU, respectively. The mixed hardwood would be a much better energy choice in this situation.

Let's take another look at using a wood product in the form of wood pellets. Wood pellets can be produced from waste resulting from other wood-processing activities. Wood pellets have a small carbon footprint. Plants convert sunlight and carbon dioxide into sugars, and oxygen is produced as a secondary product. Wood is made of these sugars and therefore can be considered a form of solar energy. Wood absorbs as much carbon dioxide as it releases when burned. Therefore, except for production and transportation, it can be considered as a carbon-neutral energy source. Note, however, that shipping a ton of wood pellets about 600 miles can use as much energy as the pellets themselves contain; anything further than that, depending on fuel prices, and you could be using more energy to transport the pellets than you will receive from burning them. Wood pellets from waste wood can be an appropriate measure for carbon-dioxide reduction, to the extent that waste wood is available. If a source of wood pellets is available at $215/ton, the equivalent energy cost versus fuel oil at $3.50/gallon would be $15.47/MBTU versus $34.65/MBTU, respectively.

Wood Pellets

$$\frac{8200 \text{ BTUs}}{\text{lb}} \times \frac{2000 \text{ lbs}}{\text{ton}} \times 0.85 \text{ (efficiency)} = 13.9 \text{ MBTUs/ton}$$

$$\frac{\$215.00}{\text{ton}} \times \frac{1 \text{ ton}}{14.3 \text{ MBTUs}} = \mathbf{\$15.47/MBTU}$$

Fuel Oil

$$\frac{138,\ 800\ \text{BTUs}}{\text{gallon}} \times 0.75\ (\text{efficiency}) = 104,\ 100\ \text{BTUs/gallon}$$

$$\frac{104,\ 025\ \text{BTUs}}{\text{gallon}} \times \frac{1\ \text{MBTU}}{1,\ 000,\ 000\ \text{BTUs}} = 0.104\ \text{MBTUs/gallon}$$

$$\frac{\$3.50}{\text{gallon}} \times \frac{1\ \text{gallon}}{0.101\ \text{MBTUs}} = \mathbf{\$34.65\ /\ MBTU}$$

Let's take a look at another example just to get a better understanding for determining energy comparisons. If the cost of electricity is \$0.14/kWh and fuel oil is available at \$3.50/gallon, another cost per unit energy comparison can be calculated as follows:

Fuel Oil

Fuel Oil at a price of **\$34.65** per MBTU (as previously calculated).

Electricity

$$\frac{3414\ \text{BTUs}}{\text{kWh}} \times 1.0\ (\text{efficiency}) = 3414\ \text{BTUs/kWh}$$

$$\frac{3414\ \text{BTUs}}{\text{kWh}} \times \frac{1\ \text{MBTU}}{1,\ 000,\ 000\ \text{BTUs}} = 0.003414\ \text{MBTUs/kWh}$$

$$\frac{\$0.14}{\text{kWh}} \times \frac{1\ \text{kWh}}{0.003414\ \text{MBTUs}} = \mathbf{\$41.01\ /\ MBTU}$$

Therefore, the equivalent energy cost of fuel oil in comparison with electricity in this example is \$34.65/MBTU versus \$41.01/MBTU for electricity, respectively. Fuel oil in this case would be a better energy choice than electricity. In this scenario, electricity would have to cost less than \$0.118 cents/kWh to compete with the \$3.50/gallon price of oil. Under most conditions, electric heat is the most expensive means to heat a dwelling unless you can produce the electricity by solar photovoltaics or wind, which would require offsetting initial costs of the systems over a long period of time. If you are heating water with an oil furnace, however, it could be less expensive to heat water with electricity because your furnace would not have to run during the warmer months. In addition, you should understand that most oil burners operate at about 80% efficiency to produce domestic hot water (DHW), not including other efficiency losses, thereby increasing actual costs. This **annual fuel utilization efficiency** (AFUE) is a measure of how efficient the furnace is as a ratio of heat output compared with the total energy consumed. An AFUE of 80% means that 80% of the energy in the fuel becomes heat for the home and the other 20% escapes up the chimney and elsewhere. The AFUE does not include the heat losses of piping or ductwork, which can be as much as 35% of the energy output of the furnace. Energy output for each situation is different for oil, propane, and gas depending on efficiencies, usage, and cost per BTU. The previous examples illustrate how to examine a level playing field as a cost per BTU for the type of fuel under consideration.

3.2 PRINCIPLES OF HEAT ENERGY

The daily BTU requirement for heating domestic water can be determined using the following mathematical relationship:

$$(\text{Amount of heat required}) \; Q = (\text{Wc}) \, (\text{Ts} - \text{Ti}) \, (\text{Cp}) \, (8.33)$$
$$\text{or}$$
$$Q = \text{Wc} \, \Delta T \text{Cp} \tag{3.1}$$

where:

Q = the amount of heat required in BTUs

Wc = Daily hot water consumption or amount of material

ΔT = (Ts − Ti) = Difference in storage temperature and inlet water temperature

and Ts = Storage temperature

and Ti = Inlet water temperature

Cp = the specific heat of water which is 1 BTU/pound/degree increase in temperature

8.33 = Conversion constant = the number of pounds in 1 gallon of water to convert Wc from the amount in gallons to a weight in pounds

We can use the algebraic relationship above by assuming no heat losses and that the specific heat is constant for a specific temperature range. Before we use this mathematical relationship in an example of energy use, let's discuss the terms from which the above heat transfer equation is derived. We will do so to provide a distinction between the terms of heat and temperature. We also will keep in mind that this book is not intended to be strictly a physics textbook, but rather is intended to provide enough information so that you can make an informed decision about whether you might benefit from the use of solar energy in your area.

Temperature is not heat energy. Temperature is a measure of the average translational energy per molecule. It is an indicator of the intensity or degree of heat stored in a body. The most commonly used measure of temperature in the United States is the Fahrenheit (°F) scale. Heat is transferred between two substances at different temperatures, always from the higher to the lower temperature. Whether an object feels hot or cold depends on the direction of the heat transfer. Objects at temperatures higher than our body temperature feel hot because heat flows from the object to our body. Conversely, objects at lower than body temperatures feel cold because heat from our body is transferred to the colder object. The following example illustrates the difference between heat and temperature. Using the algebraic equation, $Q = \text{Wc}\Delta T\text{Cp}$, we can determine what increase in heat energy is necessary to increase the temperature of 1 gallon of water from 70 °F to the boiling point at 212 °F.

$$Q = \text{Wc} \, \Delta T \, \text{Cp}$$

where:

Cp = 1 BTU/lb-°F (specific heat)

Wc = 1 gallon × 8.33 lb/gallon = 8.33 lb (water consumed in pounds)

ΔT = 212 °F − 70 °F = 142 °F (temperature increase in °F)

Therefore, Q = (1 BTU/lb-°F) (8.33 lb) (142 °F) = 1183 BTUs (amount of heat energy required)

So, to raise 1 gallon of water from 70 to 212 °F, it will take 1183 BTUs. If we want 2 gallons of water heated to the same temperature, then twice as much energy is needed to achieve the same result.

3.3 EXPLANATION OF HEAT TRANSFER

The concept of heat transfer is discussed because the efficiency of a solar DHW collector depends on the physical mechanisms by which heat may be transferred from one place to another. Different materials absorb different amounts of heat energy at a different rate of transfer to the same difference in temperature. A metal object transfers heat at a faster rate than a nonmetal object and therefore feels colder to the touch because of its greater ability to absorb heat. The measure of a material's ability to absorb heat is referred to as its *specific heat.* The **specific heat, Cp**, is the quantity of heat (in BTUs) absorbed by 1 lb of a material to produce a 1 °F temperature change (**BTU/lb-°F**). (In the metric system, it is the number of calories absorbed by 1 g of material to rise 1 °C in temperature.) The reason for the variation of specific heat from substance to substance lies in the different masses of the atoms. The ratio between the specific heats of two bodies of equal masses is defined by the ratio of the temperature changes experienced when the two bodies are brought into thermal contact with one another. As the two bodies contact each other, they approach a final equilibrium temperature, and the temperature changes are found to be inversely proportional to the respective masses of the bodies. Therefore, if the temperature changes of two bodies A and B with equal mass are found to be ΔTa and ΔTb, the ratio between the specific heats of these bodies, (Cp)a and (Cp)b, is defined as follows:

$$\frac{(Cp)\,a}{(Cp)\,b} = -\frac{\Delta\,Tb}{\Delta\,Ta}$$

Consequently, if the specific heat of water is set equal to unity as a reference, then the specific heat of other materials can be determined. Examples of fluids with varying specific heats are shown in Table 3.2.

Objects of the same material may absorb different quantities of heat energy when changing the same amount of temperature. The factor affecting a difference in the quantity of heat absorbed in this case is the *mass*. Some confusion between the concepts of weight and mass arise because the two terms routinely are interchanged in everyday language. The weight of an object is actually the measure of the force exerted on the object (mass) due to the earth's gravity.

Table 3.2 Typical Heat Transfer Liquids

	Water	50% Propylene Glycol/Water	Synthetic Hydrocarbons	Silicone
Freezing point (°F)	32	−28	−40	−58
Boiling point (°F) (at atmospheric pressure)	212	222	Up to 625	392
Fluid stability	Requires pH or inhibitor monitoring	Requires pH or inhibitor monitoring	Excellent	Excellent
Flash point (°F)	None	600	345	450
Specific heat at 100 °F (BTU/lb-°F)	1.00	0.85	0.56	0.39
Toxicity	No	No	No (typically not)	No

Source: Data are extracted from manufacturer's literature to illustrate the properties of a few types of liquid that have been used as a transfer fluid. Data are average values

Objects with greater mass have greater weight under the influence of gravity. Weight is a force and has the units of force (pounds). The mass of a body, on the other hand, is the quantity of inertia it possesses. Newton's second law states that $m = f/a$, where m is the mass, f is the force, and a is the acceleration. The constant ratio of force to acceleration therefore can be considered a property of the body called its mass. The weight of a body means that the gravitational force is exerted on it by the earth. On earth, the standard pound by definition is a body of mass equal to 0.4535924277 kg. The pound of force is the force that gives a standard pound mass an acceleration equal to the standard acceleration of earth's gravity which is 32.1740 ft/s². Now, let's once again determine what increase in heat energy is necessary to increase the temperature of 1 gallon of water from 70 °F to the boiling point of water at 212 °F. In units of the metric system, we have the same equation that we discussed previously:

$$Q = M\Delta T\, Cp$$

where:

Cp = 1 calorie/gram-°C

$M = (8.33\,\text{lb}) \times (0.454\,\text{kg/lb}) = 3.78\,\text{kg}$

$\Delta T = (212\,°F) - (70\,°F) = (100\,°C) - (21.1\,°C) = 78.9\,°C$

Therefore, $Q = (1\,\text{calorie/g-}°C)(3.78\,\text{kg})(78.9\,°C) = 298{,}084$ calories.

Using the conversion factor of 252 calories/BTU, we find this answer equivalent to 1183 BTUs. So, no matter whether we use the English system or metric conversions, the quantity of heat needed depends on the parameters of specific heat, mass or weight converted, and temperature difference. We previously mentioned the term "specific heat" to point out that different transfer fluids, whether it is water with a specific heat of 1, a propylene glycol mixture with a specific heat of 0.85, or some other fluid, each produce different results in the

Table 3.3 BTU of Heat Supplied by 1 gallon at 1 °F Temperature Rise

Heat Transfer Fluid	Specific Heat (BTU/lb-°F)	Density (lb/gallon)	No. of BTUs Supplied
Water	1.00	8.33	8.33
Water/Glycol	0.83	8.3	6.89
Hydrocarbon	0.56	7.0	3.92
Silicone	0.39	8.0	3.12

transfer of heat in a solar hot water collector. The specific heat, flow rate, and density constitute the solar energy collection capability of the transfer fluid. The type of heat transfer fluid used in a solar hot water system is therefore important in determining the amount of heat provided. The heat transfer fluids depicted in Table 3.3 will transfer heat at different capacities.

This heat transfer is a function of the specific heat multiplied by the weight transferred in a given amount of time. For example, if it takes a flow rate of 1 lb/min of water to transfer 1 BTU, then by comparison, it would take a flow rate of 1.2 lb/min of water/glycol, 1.79 lb/min of a hydrocarbon, and 2.56 lb/min of silicone to transfer the same amount of heat because of the lower specific heat.

The density (mass per unit volume) of the transfer media, or the weight/unit, also has a major effect on the energy collection. The weight per unit of liquid transferred is a function of volume and density. Therefore, pumping rates will vary for each heat transfer fluid to transfer an equivalent amount of heat. When flow is adjusted or normalized, each fluid will deliver the same amount of heat.

3.4 TYPES OF HEAT TRANSFER

There are three different physical mechanisms by which heat is transferred from one place to another. These include conduction, convection, and radiation. The following information is included to provide additional background in reference to the explanation of Solar DHW Collector Performance, which is discussed in Chapter 4, Section 4.3.

3.4.1 Conduction

Conduction is the transfer of energy through a material by direct molecular interaction. The heated molecules transfer some of their vibrational energy directly to adjacent cooler molecules, resulting in a large-scale energy transfer. Energy is lost by heat conduction through direct physical contact with objects of lower temperature. Conversely, heat is gained by direct contact with objects of higher temperatures. The ability of a material to permit the flow of heat is called its thermal conductance, C. Thermal conductance is the quantity of heat per unit time that will pass through the unit area of a particular material or body when a unit average temperature is established between the surfaces. The units of measure are BTU-in/hr-ft^2-°F. Thermal conductivity, K, specifies the thermal conductance of a material for a certain thickness, such that $K = Ct$, where C is the thermal conductance and t is the thickness of the material. The units of measure are BTU-in/hr-ft^2-°F. The rate of heat flow, ΔQ (BTUs/hr), depends on the thermal conductivity of the material, the cross-sectional area of the conductor, its thickness, and the temperature difference between the surfaces considered. The rate of heat flow is directly proportional to the area through which the heat energy can move and is inversely proportional to the thickness of the material as shown in the heat conduction Eqn (3.2).

$$\Delta Q = (K/t) A (T_2 - T_1) \tag{3.2}$$

where:

ΔQ = rate of heat transfer (BTUs/hr)
K = thermal conductivity (BTU-in/hr-ft^2-°F)
t = thickness of material (in)
A = surface area (ft^2)
$(T_2 - T_1) = (\Delta T)$ = temperature difference between surfaces (°F)

The rate of heat transfer by conduction from the back of an absorber plate in a flat-plate solar collector for DHW to the outside environment is illustrated in the following example. If a collector has a 24 ft^2 absorber plate with a 2-inch fiber insulation of thermal conductivity, K, of 0.23 BTU-in/hr-ft^2-°F, an absorber plate temperature of 180 °F and an ambient air temperature of 50 °F, the rate of heat by conduction would be as follows:

$$\Delta Q = (0.23/2) (24) (180 - 50) = 358.8 \text{ BTUs/hr}$$

The lower the K value or greater the thickness of insulating material, the lower the rate of heat transfer by conduction. (Values of thermal coefficients can be found in the ASHRAE Handbook—Fundamentals.)

The tendency of a material to retard heat transfer is known as **thermal resistance**, R. Thermal resistance of a material is the inverse of its thermal conductance such that $R = 1/C$. The units of measure are hr-ft^2-°F/BTU. All materials have some resistance to heat flow. One that has a high thermal resistance is called **insulation**. The concept of thermal resistance is useful in calculating the heat loss through a composite wall. In many situations, the same amount of heat must flow through each insulative layer. This type of situation is similar to an electrical circuit with elements connected in a series. Series circuits combine the overall circuit resistance as the sum of the individual resistances. For example, a wall composed of 0.5 in of inner wallboard ($R_1 = 0.45$), 3.5 inches of insulation ($R_2 = 11$), and 0.5 in of outside sheathing ($R_3 = 1.32$, and wood siding ($R_4 = 0.81$) would have a total overall resistance, $R_T = R_1 + R_2 + R_3 + R_4 = 13.58$ hr-ft^2-°F/BTU.

As additional thermal resistances are involved, the overall effect is simply the sum of the individual components. The overall coefficient of transmittance, U, is the reciprocal of the total thermal resistance such that $U = 1/R_T$. The units of measure are the same as thermal conductance. By definition, therefore, Eqn (3.2) can be represented as Eqn (3.3).

$$\Delta Q = UA\,(\Delta T) \tag{3.3}$$

where:

ΔQ = overall rate of heat transfer (BTUs/hr)
A = overall area (ft^2)
U = overall coefficient of transmission (BTU/hr-ft^2-°F)
ΔT = temperature difference between surfaces (°F)

Assume that the inside temperature of a house was to be maintained at 68 °F. If the outside ambient air was 40 °F and the total wall area was 256 ft^2 (8 ft × 32 ft) with an overall thermal resistance, R_T, of 13.58 ($U = 0.074$), the total rate of heat transfer, ΔQ, could be found using Eqn (3.3) as follows:

$$\Delta Q = UA\,(\Delta T)$$
$$\Delta Q = (0.074)\,(256)\,(68 - 40) \text{ and}$$
$$\Delta Q = 530.4 \text{ BTUs/hr}$$

Note that the heat loss through this wall in this particular example does not include convective losses.

3.4.2 Convection

Convection involves the transfer of heat energy by the actual movement of the heated fluid in contact with solid surfaces. The air or liquid molecules exchange energy with adjacent molecules by carrying the energy via a fluid transport. There are two types of convection: natural or free and forced. *Natural convection* or free convection occurs due to the heating or cooling of any fluid when it contacts an object. As the air changes temperature, it changes density and rises or falls due to the action of gravity. *Forced convection* occurs when the fluid has a significant velocity relative to the object encountered. A fluid at a higher speed of travel will cause more heat transfer than one at a lower speed of travel. For example, a person will feel colder on a windy day of 10 °F than on a day with the same temperature and no wind at all, because the body heat is dissipated quickly. This effect is known as the "chill factor". The rate of heat transfer due to convection is similar to the conduction Eqn (3.2). Convection is directly proportional to the temperature difference between the surface and adjacent fluid, $(T_s - T_f)$, the heat transfer area, A, and a film or surface coefficient, h, as shown in Eqn (3.4).

$$\Delta Q = Ah\,(T_s - T_f) \tag{3.4}$$

where:

ΔQ = rate of heat transfer (BTUs/hr)

A = surface area (ft^2)

h = film or surface coefficient (BTU/hr-ft^2-°F)

$(T_s - T_f)$ = temperature difference between surface and adjacent fluid (°F)

The film or surface coefficient, h, increases with fluid velocity. For instance, the free convection coefficient of air next to an inside wall (still air) is approximately 1.5 BTU/hr-ft^2-°F. Similar to the discussion of conduction, the thermal resistance, R, would translate to $R = 1/h$, resulting in 1/1.5 BTU/hr-ft^2-°F, or an $R = 0.67$. The thermal resistance for the inside wall would be $R_5 = 0.67$. On a windy day with 15 mile per hour wind speeds, the free convection coefficient of air next to an outside wall is approximately 5.9 BTU/hr-ft^2-°F, resulting in a thermal resistance $R = 0.17$. The thermal resistance for the outside wall would be $R_6 = 0.17$. (As mentioned previously, other examples of coefficients can be found in the ASHRAE Handbook—Fundamentals.) The total heat loss through the wall mentioned previously in the discussion of conduction can now be calculated to include the convective losses that also exist. The total overall thermal resistance including conduction and convection is therefore $R_T = R_1 + R_2 + R_3 + R_4 + R_5 + R_6 = 14.42$ hr-ft^2-°F/BTU.

The overall rate of heat transfer including both conduction and convection resistances can be determined using Eqn (3.3). If the inside temperature of a house was to be maintained at 68 °F, and the outside ambient air was 40 °F, and the total wall area was 256 ft^2, the total rate of heat transfer would be

$$\Delta Q = UA(\Delta T)$$
$$\Delta Q = (0.069)(256)(28) \text{ and}$$
$$\Delta Q = 494.6 \text{ BTUs/hr}$$

It can be seen from this example that the greater the velocity of the moving air, the more the resistance of the wall decreases; thus, the increase in total heat loss.

3.4.3 Radiation

Radiation involves the transfer of energy from a warm body by electromagnetic waves. Heat transferred in this way often is referred to as thermal radiation to distinguish it from electromagnetic signals. This form of heat transfer does not require a medium for propagation, and direct contact with the radiating source is not necessary. As illustrated in Chapter 2, radiant energy exists at varying wavelengths, including gamma rays, X-rays, ultraviolet rays, visible rays, infrared rays, radio waves, and electric waves. When a body absorbs radiation, the body tends to restore its original state by reradiating and redistributing the extra energy. As a result of the energy redistribution, the emitted radiation may have a wavelength distribution different from that of the originally absorbed radiation. The distribution is controlled mainly by the temperature of the body. When a heated object emits a maximum amount of radiation as efficiently as possible regardless of the emitting surface, it is called a **blackbody** emitter. The thermal radiation properties of a material are described by its overall emissivity, ϵ. By definition, *emissivity*, ϵ, is the ratio of actual power reradiated at any wavelength to the power that would be emitted by a perfect blackbody at that wavelength. We usually find that the radiation for real bodies is not distributed quite the same as that of a blackbody. Therefore, the body is assigned an overall emissivity, ϵ, such that at a certain temperature, the body emits a fraction ϵ of the energy emitted by a blackbody at that temperature. Furthermore the body is assigned the properties of reflectivity (ρ), absorptivity (α), and transmissivity (τ), which accounts for the total intensity (*Io*) incident upon a surface. The sum of these properties is equivalent to the unity as shown in Eqn (3.5).

$$(\rho)\,Io + (\alpha)\,Io + (\tau)\,Io = 1 \tag{3.5}$$

At any given temperature, the emitting power of a blackbody is directly proportional to its absorbing power, and at any given wavelength, we have $\epsilon(\lambda)$ equivalent to $\alpha(\lambda)$. Note, therefore, that the properties ϵ, ρ, α, and τ lie between 0 and 1 for real bodies. For a true blackbody these values would be 1, 0, 1, and 0, respectively. The ideal absorber plate has a surface with a high absorptivity to absorb as much solar radiation as possible and a low emissivity to reduce the thermal reradiative losses. Such an absorber is said to have a *selective surface*.

3.5 CALCULATING HOT WATER AND ELECTRICAL ENERGY REQUIREMENTS

To determine the energy worth of a solar DHW or photovoltaic system, we must first determine how much energy we need to either heat water or to provide electricity for our monthly demands. Once we know what our heat and electricity requirements are, we then need to determine the amount of solar radiation that is available. We then can appropriately size the collector system for either hot water or electricity applications as discussed in Chapter 4, Section 4.6, and Chapter 5, Section 5.4, respectively.

3.5.1 Hot Water Requirements

To determine the number of solar hot water panels required to meet your hot water energy requirements, you first need to determine the quantity of hot water usage and the resulting energy requirements. In other words, how many BTUs will the system need to produce to satisfy our demands?

The amount of hot water we need each day depends on the number of occupants in the dwelling. The usage will vary from residence to residence and depends on the amount of laundry, dishwashing, and personal hygiene as depicted in Table 3.4. The hot water consumption table illustrates estimates of hot water usage. These estimates can be further refined by referring to your particular appliance's specification sheet, but such data refinement is not considered necessary for residential applications.

Table 3.4 Hot Water Consumption Table

Clothes Washing Machine	Gallons per Use of Hot Water 14 lb Machine	18 lb Machine	Electrical Use kWh/Load
Hot wash/warm rinse	28 gallons	36 gallons	4.5 kWh
Hot wash/cold rinse	19 gallons	24 gallons	2.8 kWh
Warm wash/cold rinse	10 gallons	12 gallons	1.9 kWh
Dishwashing	Small	Large	
Dishwashing machine	4–6 gallons	10 gallons	0.87–1.59 kWh
Sink washing	4–8 gallons	N/A	N/A
Personal hygiene			
Tub bathing	12–30 gallons		
Wet shaving/hair washing	2–4 gallons		
Showering	2–6 gallons/min		

Based on the Table 3.4, you can determine the daily household BTU requirement by multiplying the daily hot water consumption, Wc, in gallons by 8.33 pounds of water/gallon to obtain the number of pounds of water, then by multiplying that number by the water's specific heat, Cp, (1 BTU/pound/degree increase in temperature needed), and then by multiplying that result by the average temperature increase (Ts−Ti) desired for hot water. For instance, assume a hot water storage tank is used and the storage temperature, Ts, is maintained at approximately 135 °F. Also assume that the inlet temperature, Ti, from a well or city water, enters storage at 40 °F in the winter and 50 °F in the summer. The average temperature increase, Ts−Ti, in a winter condition scenario would be 95 °F (135 °F–40 °F).

Let's assume that we have determined that a typical dwelling of four would require 70 gallons of hot water/day. Remembering the previous Heat Equation and assuming a winter season scenario, the following conditions would exist:

$$\text{Heat Equation: } Q = Wc\,\Delta TCp$$

where:

(Cp) = 1 BTU/lb-°F

$\Delta T = (Ts - Ti) = 135 - 40\,°F = 95\,°F$

(Wc) = 70 gallons of water (1 gallon of water equals 8.33 lbs)

Using this mathematical relationship, our daily hot water BTU requirement would be as follows:

$$(Wc)\,(Ts - Ti)\,(Cp)\,8.33 = \left(\frac{70\ \text{gallon}}{\text{day}}\right) \times \left(\frac{8.33\ \text{lb}}{\text{gallon}}\right) \times (95\ °F) \times \left(\frac{1\ \text{BTU}}{\text{lb-}°F}\right)$$

$$= 55,395\ \text{BTUs/day}$$

From this example, we now know how much energy we need to supply our daily hot water requirements. Before we can determine how many solar collectors we need to meet that demand, we must next determine the amount of solar insolation that is available in your geographic location. Note that the term *insolation* is not the same as the term *insulation*. *Insolation* is defined as the total amount of solar energy received at the earth's surface at any location and time (**BTU/ft²-hr**), whereas insulation is a material that has a high thermal resistance to heat flow. We will discuss the amount of solar radiation available and the method used to size a solar DHW system in Chapter 4, Sections 4.5 and 4.6, respectively. Knowing the demand requirements and the amount of energy available, you can easily calculate the size of the Solar DHW system using Table 4.3 and the worksheet in Table 4.4a, provided in Chapter 4, Section 4.5.

3.5.2 Electricity Requirements

To determine the number of solar photovoltaic panels required to meet our electricity requirements, we need to determine the amount of power we consume in terms of kilowatt-hours. In other words, how many kWh will the system need to produce? This information can be acquired by reviewing your monthly electric bills. Simply collect your monthly electrical bills from the past 12 months (or contact your utility company for historical records), add up the total usage in kWh for the entire year, and divide the total by 365 days in the year to determine your average kWh usage per day. For instance, if you use 800 kWh of electricity/month for a total of 10,800 kWh annually, then the daily electrical demand would be 29.6 kWh/day:

$$\frac{10,800 \text{ kWh}}{365 \text{ days}} = 29.6 \text{ kWh/day}$$

Knowing the electrical demand requirements, we then can calculate the number of solar photovoltaic modules necessary to provide that amount of electricity by using the information and method provided in Chapter 5, Section 5.4.

Chapter four

Solar Domestic Hot Water Systems

4.1 MAIN SYSTEM TYPES

Many manufactured solar domestic hot water (DHW) systems are available to the homeowner for heating domestic water. The choice can be confusing, however, if one is unfamiliar with the basic function of each type under consideration. Before making your initial investment, you should determine which system type would best serve your needs.

There are two generic classes of active solar DHW systems. These include concentrating and nonconcentrating collection systems. Evacuated tube-type collectors are categorized as concentrating collectors and offer a high British thermal unit (BTU) yield under hazy, diffused sunlight conditions; however, they can be more costly than nonconcentrating collectors, otherwise known as flat-plate collectors. Because there are ongoing improvements with each type, you should discuss costs and comparisons at that time with a certified solar dealer before making a decision.

The three most prevalent types of solar DHW systems include (1) closed loop freeze resistant, (2) drain back, and (3) drain down. Each of these types are categorized as either closed-loop (indirect) or open-loop (direct) systems.

4.1.1 Closed-Loop Freeze-Resistant System

The most common type of closed-loop (indirect) system used is the closed-loop freeze-resistant system, as illustrated in Figure 4.1. This type of system heats a freeze-resistant transfer fluid, which in turn heats the domestic water through a heat exchanger in the storage tank. The freeze-resistant fluid heated in the collectors is actually the secondary fluid. The heat from this secondary fluid is passed to the primary fluid (tap water from the storage tank) through the heat exchanger, such as a finned copper coil. There is no exposure of the fluid to the atmosphere, and the system is pressurized. Because there are only friction losses to overcome, a small circulator pump (typically 1/20th horsepower (hp)) is normally used with 0.75-inch diameter copper piping, depending on the length of run. A nontoxic transfer fluid is circulated through this closed loop whenever sufficient insolation is available. Temperature control is established with a differential controller. This device signals the circulator to start when there is a

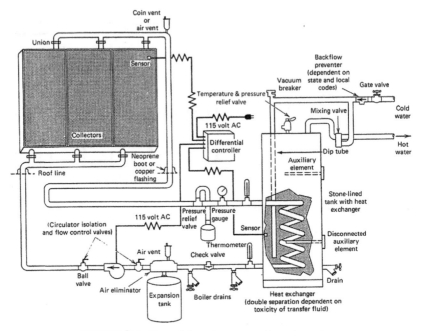

FIGURE 4.1 Closed-loop freeze-resistant system.

sufficient temperature gradient (15–20 °F) between the collectors and storage tank so heat can be accumulated. It also signals the circulator to stop when the storage tank temperature is within 3–5 °F of the collector fluid outlet so heat will not be lost from storage. This is a reliable type of system to use because, with the proper heat transfer fluid, you will not damage the system during freezing conditions.

Components necessary for a closed-loop freeze-resistant system as shown in Figure 4.1 include the following:

Solar collectors with associated union joints	Pressure gauge (1)
1/20 horsepower circulator with (2) ball valves for isolation	Temperature gauges (2)
Expansion tank	Air vent or coin vent
Storage tank with exchanger (stoned lined tank preferable)	Differential controller and sensors
Fill-drain assembly (includes two boiler drains and a check valve)	Back-flow preventer (depending on local plumbing codes)
Check valve	Vacuum relief valve
Air purger with air vent	Pressure relief valve

A typical plumbing hardware arrangement with respect to this type of system is illustrated in Figure 4.2.

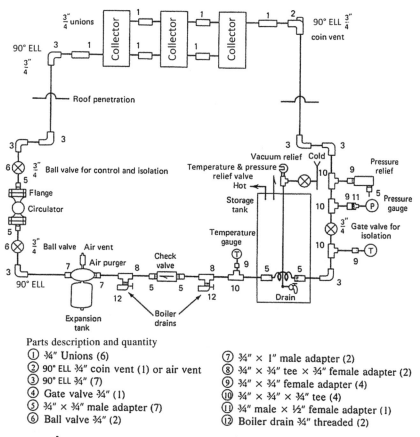

Parts description and quantity

① ¾" Unions (6)
② 90° ELL ¾" coin vent (1) or air vent
③ 90° ELL ¾" (7)
④ Gate valve ¾" (1)
⑤ ¾" × ¾" male adapter (7)
⑥ Ball valve ¾" (2)

⑦ ¾" × 1" male adapter (2)
⑧ ¾" × ¾" tee × ¾" female adapter (2)
⑨ ¾" × ¾" female adapter (4)
⑩ ¾" × ¾" × ¾" tee (4)
⑪ ¾" male × ½" female adapter (1)
⑫ Boiler drain ¾" threaded (2)

FIGURE 4.2 Miscellaneous plumbing hardware (closed-loop freeze-resistant system).

4.1.2 Drain-Back System

The second most widely used system is the drain-back system. This type of system provides passive freeze protection without a freeze-resistant fluid, and it can be used in tandem with a glass-lined hot water storage tank. A drain-back system differs from that of a closed-loop system in that normally city or well water flows through a heat exchanger instead of a freeze-resistant transfer fluid. The domestic water supply remains in a closed loop as illustrated in Figure 4.3, and the collector loop remains open and unpressurized. Water remains in the collector loop only while the pump is running. When the temperature difference is not adequate to provide heat to storage, a differential controller shuts off the pump, and the water drains automatically into the storage tank by gravity. One must ensure that the internal or external manifolds of the collectors are pitched at least 1 in/3 ft so proper drain back will occur. Water is used in the collector loop instead of freeze-resistant fluid because it would not be economical to fill the storage tank with a freeze-resistant fluid. Furthermore, the specific heat of water is superior to any of the transfer fluids. Storage tanks for drain back systems are

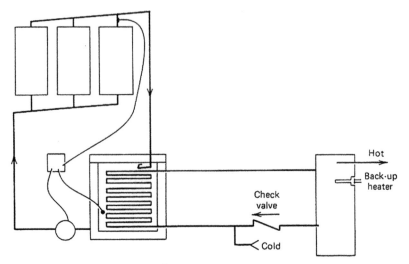

FIGURE 4.3 Drain-back system.

typically 90-, 120-, or 150-gallon insulated polyethylene containers depending on whether a three-, four-, or five-panel system is needed, requiring additional fluid containment. The storage tank is unpressurized (open-loop), and supply and return lines connect the tank and collector. A pump rather than a circulator is used and is sized to overcome the static head. A one-seventh horsepower pump normally will provide the 40- to 50-foot head required. System efficiency can be enhanced somewhat if a return path is established with a check valve to the cold water inlet of the storage tank. This provides a thermosiphoning arrangement between the back-up heater and storage tank and therefore increases heat storage capacity. The check valve must be tilted so that its gate is vertical, allowing valve operation by the small hydrostatic force caused by thermosiphoning.

Another type of drain-back system also is available called SECUSOL® that is actually a closed-loop hybrid solar thermal system with an integrated drain-back function using proprietary technology that eliminates stagnation or freezing. The storage tank contains an integrated drainage reservoir with a heat exchanger and receives all fluid from the collectors and solar piping loop when the solar pump is idle. The maximum height of the system of such a system is approximately 28 ft and the required piping is 12 mm (approximately 0.5 in) in diameter. The storage tank includes the differential controller and pump and does not require an expansion tank, reducing installation and maintenance. Such a system normally uses a 30% mixture of propylene glycol with water as a transfer fluid. When solar radiation is available and the storage tank requires heat, the pump is activated automatically by the differential controller and circulates the fluid in the heat exchanger replacing the air in the collectors, which then is forced into a thin layer inside the heat exchanger. When the pumps stops, the fluid from the panels replaces the air, which is cycled back into the collectors, protecting them from freeze conditions.

4.1.3 Drain Down System

The drain-down system is a closed-loop system and differs from the closed-loop freeze-resistant system and drain-back systems in that no heat exchanger is used, thereby increasing effective heat transfer. Potable water is circulated directly from the storage tank through the collector loop. Freeze protection is provided by a differential controller, which deenergizes three solenoid valves to their normally open or closed positions when the ambient temperature approaches 32 °F. This type of system is illustrated in Figure 4.4.

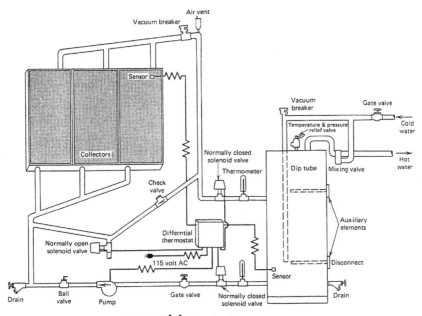

FIGURE 4.4 Drain-down system.

When sufficient insolation is available, the solenoid valves are actuated, the drain-down loop is closed, and the potable water loop from storage to collectors is opened. Collector manifolds and piping must be pitched so the system will automatically drain down upon solenoid deactivation. Major components necessary for a drain down system include the following:

Solar collectors with associated union joints	Pump (bronze or stainless steel)
Storage tank with vacuum relief valve	Solenoid drain valves
Check valve	Temperature/pressure relief valve
Air vent	Temperature gauge and miscellaneous gate/ball valves

4.2 BASIC SYSTEM CONFIGURATION

The collector area is composed of individual collectors arranged to operate as a single system. The arrangement and relationship of one collector to another is extremely important for effective solar collection and efficient system operation.

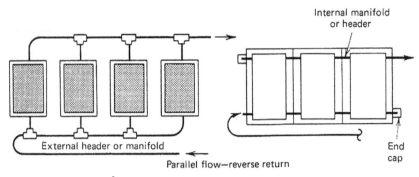

FIGURE 4.5 Parallel flow-reverse return arrangement.

Reverse return piping systems are preferred because the total length of supply piping and return piping serving each collector is the same, the pressure drop across each collector is equal, and the pressure drop across each manifold is also theoretically equal. The transfer fluid flow through each collector is relatively the same. External and internal manifold arrangements are illustrated in Figure 4.5.

4.3 SOLAR DHW COLLECTOR PERFORMANCE

4.3.1 Heat Energy Collection and the Solar Collector

A solar DHW collector is designed to collect both diffuse and direct beam radiation while maintaining minimum heat loss. The principal heat loss factors include (1) conduction loss from the back of the absorber plate through the insulating material, (2) conduction losses through the sides of the collector, (3) convection losses upward through the glazing, and (4) the upward radiation loss. These heat losses can be quite large because the area for such losses is essentially equal to the area of energy collection.

Figure 4.6 illustrates the heat loss of a flat-plate collector without a glazing cover. Much of the radiation absorbed by the flat black absorber plate is lost from the top surface because of convection and radiation. Convection losses can exceed radiation losses by more than a factor of five at wind speeds of only 10 miles per hour. Useful heat is retained without a glazing cover only if the temperature of the absorber plate is close to the temperature of the ambient air.

Figure 4.7 illustrates the heat loss of a collector with a glazing cover. In this situation, the radiation absorbed by the flat black absorber plate is reemitted, but the glazing cover blocks much of the loss of this reemitted radiation to the outside. Energy is trapped in two ways. The temperature of the reradiated energy from the absorber surface is such that the energy distribution curve (shown previously as Figure 2.5) is shifted toward the right so the surface only emits infrared (long-wave) radiation. A solar collector glazing is essentially opaque to long-wave radiation and reradiates this energy back to the absorber plate. The glazing also traps a layer of still air next to the absorber and reduces the convection heat loss. This combination of energy entrapment is a phenomenon

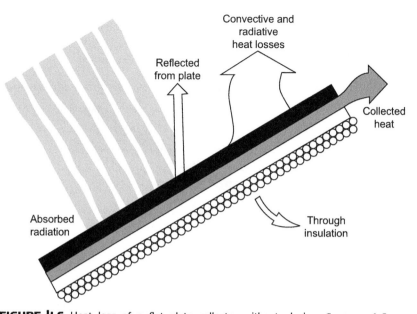

FIGURE 4.6 Heat loss of a flat plate collector without glazing. Courtesy of Copper Development Association, Inc. New York, New York.

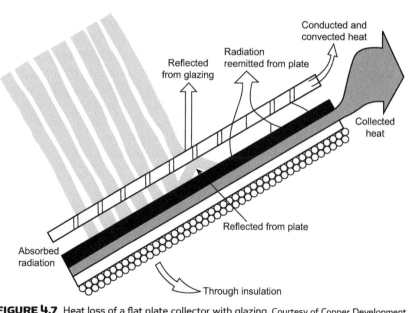

FIGURE 4.7 Heat loss of a flat plate collector with glazing. Courtesy of Copper Development Association, Inc. New York, New York.

known as the "greenhouse effect". There is some heat loss through conduction and convection as illustrated. A reflected energy loss is established with the addition of a glazing cover. If glass is used as a glazing cover, it can be etched by a thin film, such as a fluoride-based acid bath, so that the overall reflective

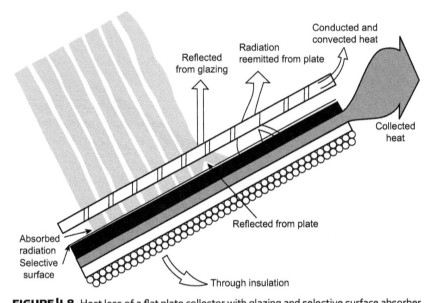

FIGURE 4.8 Heat loss of a flat plate collector with glazing and selective surface absorber. Courtesy of Copper Development Association, Inc. New York, New York.

loss is reduced. The amount of absorption in the glass also can be reduced by lowering the iron content.

Figure 4.8 illustrates the heat loss of a collector with a glazing cover and an absorber with a selective surface. The selective surface reradiates a much smaller portion of the absorbed energy than does a flat, black nonselective surface. There is still some heat loss through conduction and convection to the outside air. This type of surface is sensitive to contamination by dust and does not retain its unique properties if exposed to the weather.

4.3.2 Determining Collector Efficiency

The following information is a little more technical than other portions of this book and has been included if you wish to better understand the parameters used to determine solar hot water collector efficiency. If you want to avoid the algebra and graphical derivations that explain collector efficiency, you can skip this portion and go directly to the examples at the end of this section. Once you have received information regarding the solar collector's specifications available from your local dealers, you will be able to compare the different manufacturers to determine which collector would work best for your particular circumstances. In addition to collector *efficiency* ratings, you also can compare collector *certification* ratings that are available from the Solar Rating and Certification Corporation (SRCC) (www.solar-rating.org). SRCC ratings are discussed later in this Chapter.

Solar collectors normally are tested with American Society of Heating, Refrigeration, and Air Conditioning Engineers (ASHRAE) test method 93-86 (a revision of ASHRAE Standard 93-77) and ISO 9806-1,3 or the European Standard EN 12975-2. Both American and European testing standards are comparable with one another. The European test standards allow for a wider range of test conditions, but at the same time, these standards are comparable to the steady-state methods of the older ASHRAE 93-77 test methods. These test methods provide efficiency versus operating conditions to construct a normalized curve for insolation, I_0, and temperature difference between the heat transfer fluid collector inlet temperature, Ti, and ambient air, Ta. A typical collector efficiency curve is illustrated in Figure 4.9. This type of graph should be available for all

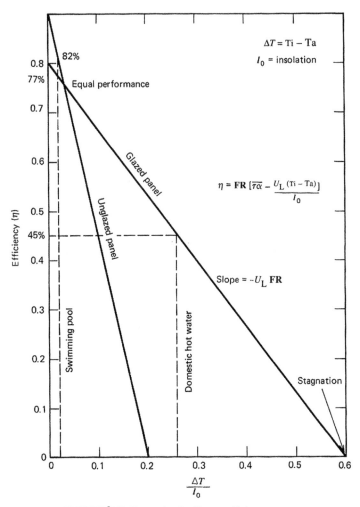

FIGURE 4.9 Example of collector efficiency curve.

manufacturer's data. Because all data should be derived under the same testing requirements, all graphs should be comparable with one another. Although the graphic illustration may look a bit foreboding at first, it is important to understand how this curve is derived and what it actually means.

The simplest type of collector consists of a flat black plate without any glazing. The heat to be extracted from this plate will be less than the solar energy collected by the plate because of convection or conduction and radiation loss of the collector to the surroundings. This amount of heat loss is a function of the temperature difference between the surface and its surroundings, and what is called the emissivity of the absorber plate. Efficiency, η, of the collector will be defined as output divided by input. Output is the amount of energy absorbed less losses, and input is the amount of insolation, I_o, available incident to the collector surface.

Efficiency is defined by Eqn (4.1):

$$(\text{Efficiency}) \, \eta = \frac{\text{output}}{\text{input}} = \frac{\text{absorbed energy} - \text{losses}}{\text{insolation} \, (I_o)} \tag{4.1}$$

and algebraically separating these terms,

$$\eta = \frac{\text{absorbed energy}}{\text{insolation} \, (I_o)} - \frac{\text{losses}}{\text{insolation} \, (I_o)}$$

The maximum efficiency that can be obtained by the collector would result if there were no losses as shown in Eqn (4.2):

$$\eta = \frac{\text{absorbed energy}}{\text{insolation} \, (I_o)} = \alpha \text{ where losses} = 0 \tag{4.2}$$

This ratio is designated as solar absorptance, α, which is characteristic of the collection surface. The efficiency of the collector will not exist without losses to the absorbed heat. To keep it algebraically simple, therefore, these losses are approximated as being proportional to the temperature difference between the absorber plate surface temperature, Tp, and the ambient air temperature, Ta. The convection, conduction, and radiation losses are combined into one proportional constant, U_L, which we will call the heat-loss coefficient.

Therefore, we have Eqn (4.3), which depicts the losses as a ratio of the heat-loss coefficient and the temperature difference between the absorber plate surface temperature and the ambient air.

$$\text{Losses} = \frac{U_L}{(\text{Tp} - \text{Ta})} \tag{4.3}$$

Combining Eqns (4.1–4.3), results in Eqn (4.4):

$$\eta = \alpha - U_L \frac{(\text{Tp} - \text{Ta})}{I_o} \tag{4.4}$$

Remember, we are still discussing the flat black plate only; no glazing cover sheet. Note in Eqn (4.4) that the losses are reduced as the collector operates near the ambient temperature. When, or if, these losses equal the absorbed energy, we note that the efficiency is zero. In other words:

$$\eta = 0 \text{ when } \alpha = U_L \frac{(Tp - Ta)}{I_o}$$

Once the plate temperature, Tp, can no longer increase, stagnation occurs. This is one of the most severe conditions for a solar DHW collector. It can occur in the heat of summer under no flow conditions, or on a cold, windy, clear day where convection and radiation losses are large and the heat loss coefficient, U_L, approaches the magnitude of the energy absorbed.

The simple flat-black collector without a glazing has good low-temperature applications, such as in swimming pools in the summer. For swimming pools in which the maximum pool temperature is 90 °F with an average of 80 °F, the unglazed collector will collect more solar energy. An example of this will be illustrated later in this chapter. When the difference in plate temperature and air temperature must be in excess of 100 °F, as in heating water in the winter, the value of U_L, which is the proportional constant for heat losses, becomes important. On a cold, windy, clear day, convection and radiation losses could be large such that the collector would reach stagnation, supplying no useful amount of heat. A change in the collector can be affected by either solar absorptance, α, or heat loss, U_L. Because α is typically 0.8–1.0, its effective change is less adaptable than is U_L. By decreasing the radiative and convective losses, while still maintaining a high solar absorptance, a collector can attain a greater useful amount of heat. A transparent cover or glazing therefore is used to decrease U_L. By adding a glazing, the solar absorptance is also decreased, but only slightly. This is due to a reduction in the energy transmitted to the collector plate. That transmitted fraction is designated as follows:

$$\text{``}\tau\text{''}, \text{ where}: \tau = \frac{\text{transmitted energy}}{\text{incident energy}}$$

When the collector plate absorbs this transmitted solar energy, it converts it to heat, and a portion is reradiated toward the glazing at infrared (IR) wavelengths. The IR transmission factor of most glazings is very low and therefore very little energy is reradiated to the sky. Radiation losses are reduced by trapping the radiation. Convection and conduction losses are reduced by the layer of dead air between the glazing and absorber plate.

The solar absorptance is slightly decreased because of transmission properties of the glazing, resulting in Eqn (4.5). Note the difference from Eqn (4.4).

$$\eta = \tau\alpha - U_L \frac{(Tp - Ta)}{I_o} \tag{4.5}$$

Because a portion of the energy not absorbed by the plate is reflected back to the plate by the glazing, the solar absorptance factor is an effective transmission absorptance product yielding $\overline{\tau\alpha}$. Eqn (4.5) is now written as Eqn (4.6):

$$\eta = \overline{\tau\alpha} - U_L \frac{(Tp - Ta)}{I_o} \tag{4.6}$$

Therefore, as the ambient air temperature, Ta, approaches the absorber plate surface temperature, Tp, Eqn (4.6) approaches maximum efficiency as follows:

$$\eta = \overline{\tau\alpha}$$

The plate temperature, Tp, varies continuously over the surface of the absorber plate. From a measurement standpoint, therefore, it is more convenient to measure the collector inlet, Ti, temperature than to measure the internal varying plate temperature. One might think that a better representative temperature would be the average of the inlet temperature and the outlet temperature (Ti+To)/2. Because we are more concerned with the storage outlet temperature to the collectors, however, the value of Ti would be more representative of collector efficiency.

For the efficiency to represent the entire collector, and allowing the substitution of Ti for Tp, a heat removal factor **FR** must be used to account for the fluid flow rate, collector-to-fluid interface, and inherent properties of the transfer fluid itself. Multiplying both the transmission absorptance product and the proportionate heat loss by the heat removal factor **FR**, we have Eqn (4.7).

$$\eta = \mathbf{FR}\left[\overline{\tau\alpha} - U_L \frac{Ti - Ta}{I_o}\right] \tag{4.7}$$

where

η = collector efficiency,
FR = heat removal factor,
$\overline{\tau\alpha}$ = effective transmissivity–absorptivity product,
U_L = overall heat loss coefficient,
Ti = transfer fluid temperature at the collector inlet,
Ta = ambient air temperature, and
I_o = instantaneous level of solar radiation (insolation).

Equation 4.7 represents collector efficiency and provides us with a comparable means of evaluating collector performance. You will find that comparing photovoltaic modules is a much simpler process.

Further analysis of Eqn (4.7), leads us to note that this equation is in the same format as that of the equation of a straight line where $Y = mx + b$, where b is the y-axis intercept (or ordinate) and m is the slope. From the equation of a straight line and Eqn (4.7), it can be seen that a straight line will result as illustrated in Figure 4.9, if a plot of efficiency, η, versus the quantity $(Ti - Ta)/I_o$ is made, assuming slope and intercept functions are constant. The slope of the line is a function of the overall heat loss coefficient, where $M = -\mathbf{FR}U_L$. The intercept of the line is a function of the transmissivity of the cover plate(s) and the absorptivity of the absorber plate(s), where $b = \mathbf{FR}\,\overline{\tau\alpha}$. In reality, however, U_L, is not constant (under the ASHRAE 93-77 test conditions) because it varies with the temperature of the collector and ambient air. A second-order curve therefore is used to describe the thermal performance of the collector, where

$$Y = c + bx + ax^2$$

The intercept is still related to $\overline{\tau\alpha}$ and the slope at any point on the curve is proportional to the heat loss rate for that value of $(Ti - Ta)/I_0$. From the graph of Figure 4.9, we can see that at a rate of $\mathbf{FR}U_L$, the efficiency decreases as $(Ti - Ta)/I_0$ increases and once the total absorbed energy equals the total losses,

$$\mathbf{FR}\,\overline{\tau\alpha} = \mathbf{FR}\left[U_L\frac{(Ti - Ta)}{I_o}\right] \text{ then } \eta = 0 \text{ and stagnation occurs.}$$

The efficiency curve does not give absolute values for the overall heat loss coefficient, U_L, and effective transmissivity–absorptivity product, $\overline{\tau\alpha}$, because both are multiplied by the heat removal factor, $\mathbf{FR}$. However, the plot does indicate relative values for these two quantities that can be used for comparing collectors. Determination of the absolute values of $\overline{\tau\alpha}$ and U_L would require additional measurements beyond the normal tests conducted for collector performance evaluation. So how do we compare collector performance based on efficiency curves? Which collector curve is better and for which application?

4.3.3 Examples of Using a Collector Efficiency Curve

Let's discuss two examples comparing an unglazed collector with a glazed collector. The first example concerns the use of solar radiation to heat an outdoor swimming pool; the second example to heating domestic water.

Example 1: Let's say we want to heat a swimming pool. Assuming a maximum instantaneous insolation, I_o, of 250 BTUs/ft^2-hr, an inlet water temperature of 80 °F, and an ambient temperature of 75 °F, then

$$\frac{Ti - Ta}{I_o} = \frac{80\,°F - 75\,°F}{250} = \frac{5}{250} = 0.02 = \frac{\Delta T}{I_o}$$

Efficiency, η, of the glazed collector as shown in Figure 4.9, is approximately 77% and the efficiency of the unglazed collector is 82%. The 5% difference in efficiency of these two collectors for this application equates to 12.5 BTUs/ft^2-hr [$(0.05) \times (250\,BTUs/ft^2$-hr)] more energy collected for the unglazed panel than for the glazed panel.

Example 2: Now, let's say that we want to heat domestic water. Again assuming a maximum instantaneous insolation of 250 BTUs/ft^2-hr, an inlet water temperature from storage of 140 °F and an ambient average temperature of 75 °F, then

$$\frac{Ti - Ta}{I_o} = \frac{140\,°F - 75\,°F}{250} = \frac{65}{250} = 0.26 = \frac{\Delta T}{I_o}$$

From Figure 4.9, we find the efficiency of the glazed collector is 45%, whereas an unglazed collector simply would not work in this situation. You therefore can take any collector efficiency curve and calculate the efficiency of that collector under the conditions for which you are considering its use.

4.4 BASIC SOLAR DHW SYSTEM COMPONENTS

Several components make up a complete solar DHW system. Besides the standard plumbing parts including valves, copper pipe, and other miscellaneous hardware, the main components include the solar collector, pumps, differential controllers, heat exchangers, and storage tanks. Some systems are installed with individual component parts as shown in Figures 4.1, 4.2, and 4.4 while some manufacturers have modules containing several of the main system components, including the pump, heat exchanger, and differential controller, saving installation time.

4.4.1 Solar Collectors

Evacuated tube and flat-plate collectors are the two basic types of solar DHW collectors used to transfer solar energy to heat water via a transfer medium for residential use. Because some flat plate panels have selective surface absorber plates, which provide equivalent output performance with respect to evacuated tube type collectors, the selection of which type used is primarily based on roof angle and possible snow conditions. If a roof is relatively flat when snow conditions exist, an installer might recommend the evacuated tube type if the collectors are to be raised onto a tilted rack, away from snow loads that tend to remain between tubes and along their base. The flat-plate collectors usually are preferred if the panels are to be positioned on a parallel or slightly raised tilt along the existing roof angle. Both types of panels will perform adequately and can be compared directly by reviewing their efficiency and energy output ratings.

Absorptive coatings for flat plate collectors are either selective or nonselective. Selective surfaces take advantage of differing wavelengths of solar radiation and the resulting emissive radiation from the absorbing surface. Collector performance can be increased dramatically with an absorber plate having a selective surface coating. A selective surface coating, such as black chrome, impedes the reradiation of infrared energy from the hot absorber plate, thereby retaining more heat for transfer to the liquid transfer media. The absorptivity is high and emissivity is low. A nonselective coating such as flat black paint, on the other hand, exhibits high absorptivity along with high emissivity. The higher the heating requirement, the more a selective surface is needed. Figure 4.10 is an example of a typical flush-mounted flat plate solar DHW collector system.

FIGURE 4.10 Wagner solar DHW flat plate collector array. Photo courtesy of ReVision Energy Corp. (For color version of this figure, the reader is referred to the online version of this book.)

4.4.2 SRCC Rating

The SRCC is an independent third-party certification organization that administers national certification and rating programs for solar energy equipment based on ASHRAE performance standards, as mentioned previously in the Solar DHW Collector Performance section of this chapter. The SRCC rating program is administered under SRCC document OG-100 (Operating Guidelines for Certifying Solar Collectors) and provides a means for evaluating and comparing the thermal performance of solar collectors under prescribed rating conditions. Because the performance of a solar collector will vary depending on the amount of insolation, collector tilt, collector orientation, and ambient air temperature, the purpose of the rating is to show a consumer how two or more panels would perform under a given set of conditions. The SRCC ratings, therefore, are similar to energy-efficiency ratings as applied to household appliances and automobile mileage ratings. The information contained in the SRCC rating indicate the number of BTUs the solar panel will collect compared with other models. The SRCC also evaluates the collector design and material for reliability and durability. A typical SRCC rating sheet is illustrated in Figure 4.11. You can easily compare one collector to another by determining the energy output for each dollar spent. In other words, how many BTUs does a dollar buy if spent on collector #1 versus collector #2. For example, collector #1, shown in Figure 4.11, used to heat water in a cold climate (designated as category D) on a clear day versus a cloudy day, has an SRCC rating of 23,600 BTUs/ft^2-day. If the collector cost \$400, then that collector has a dollar per energy output as follows:

$$23,600 \text{ BTU/ft}^2 - \text{day} \div \$400 = \$59/\text{BTU}$$

If collector #2 has an SRCC rating of 30,700 BTUs/ft^2-day under the same conditions with a cost of \$500, then that collector has a dollar per energy output as follows:

$$30,700 \text{ BTU/ft}^2 - \text{day} \div \$500 = \$61.40/\text{BTU}$$

SOLAR COLLECTOR CERTIFICATION AND RATING SRCC OG-100	CERTIFIED SOLAR COLLECTOR
	SUPPLIER: **Wagner Solar Inc.** 485 Massachusetts Ave, Suite 300 Cambridge, MA 02139 USA
	MODEL: **Wagner EURO C20 AR-M**
	COLLECTOR TYPE: Glazed Flat-Plate
	CERTIFICATION#: 2010035A
	Original Certification Date: 17-MAY-10

COLLECTOR THERMAL PERFORMANCE RATING

	Kilowatt-hours Per Panel Per Day				Thousands of BTU Per Panel Per Day		
CATEGORY (Ti-Ta)	CLEAR DAY (6.3 kWh / m^2.day)	MILDLY CLOUDY DAY (4.7 kWh / m^2.day)	CLOUDY DAY (3.1 kWh / m^2.day)	CATEGORY (Ti-Ta)	CLEAR DAY (2000 Btu / ft^2.day)	MILDLY CLOUDY DAY (1500 Btu / ft^2.day)	CLOUDY DAY (1000 Btu / ft^2.day)
A (-5 °C)	11.8	8.9	6.0	A (-9 °F)	40.2	30.3	20.5
B (5 °C)	10.9	8.0	5.1	B (9 °F)	37.2	27.3	17.5
C (20 °C)	9.5	6.7	3.9	C (36 °F)	32.4	22.7	13.2
D (50 °C)	6.9	4.1	1.6	D (90 °F)	23.6	14.0	5.5
E (80 °C)	4.3	1.9	0.1	E (144 °F)	14.7	6.3	0.3

A- Pool Heating (Warm Climate) B- Pool Heating (Cool Climate) C- Water Heating (Warm Climate) D- Water Heating (Cool Climate) E- Air Conditioning

COLLECTOR SPECIFICATIONS

Gross Area:	2.610 m^2 28.09 ft^2	Net Aperture Area:	2.36 m^2 25.40 ft^2
Dry Weight:	48.0 kg 106. lb	Fluid Capacity:	1.3 liter 0.3 gal
Test Pressure:	1500. KPa 218. psig		

COLLECTOR MATERIALS

Frame:	Aluminum
Cover (Outer):	Low Iron Glass
Cover (Inner):	

Pressure Drop

Flow		ΔP	
ml/s	gpm	Pa	in H2O
20.00	0.32	96.76	0.39
50.00	0.79	447.5	1.8
80.00	1.27	1045.04	4.20

Absorber Material:	Tube - Copper / Plate - Copper	Insulation Side:	Mineral Wool
Absorber Coating:	Selctive	Insulation Back:	Mineral Wool

TECHNICAL INFORMATION

Efficiency Equation [NOTE: Based on gross area and (P)=Ti-Ta]

				Y INTERCEPT	SLOPE
SI Units:	η= 0.741	-3.01900 (P)/I	-0.00871 (P)^2/I	0.747	-3.646 W/m^2.°C
IP Units:	η= 0.741	-0.53180 (P)/I	-0.00085 (P)^2/I	0.747	-0.642 Btu/hr.ft^2.°F

Incident Angle Modifier [(S)=1/cosθ - 1, 0°<θ<=60°]

			Test Fluid:	Water
Kτα = 1	-0.112 (S)	0.002 (S)^2	Test Flow Rate:	12.8 ml /s.m^2 0.0188 gpm/ft^2
Kτα = 1	-0.11 (S)	Linear Fit		

FIGURE 4.11 Typical SRCC rating sheet (Example: collector #1, category D (Ti·Ta)). Solar Rating & Certification Corporation, March 2012.

Collector #1, therefore, would be a better value for the particular conditions chosen as long as the quality of the collectors remains equal.

4.4.3 Differential Controllers

The varying nature of solar energy dictates the use of a differential type rather than a fixed type of temperature controller for all generic types of solar DHW systems. A differential controller constantly monitors and compares the temperature difference between collector temperature with storage tank temperature. When the collectors are hotter than the water in the storage tank by a preselected difference (normally 15–20 °F), a circulator-pump is activated to transfer the collected energy via a transfer medium through the open- or closed-loop system to storage. Once the storage tank has accumulated a sufficient amount of energy such that the temperature difference is within a preselected difference (normally 3–5 °F), the circulator-pump is deactivated so that the stored heat is not expended from storage back through the collectors, thus ensuring an overall net energy gain.

The sensors most commonly used to measure the temperature at the collectors and storage tank are called thermistors. These devices are made from a semiconductor material that experiences a nonlinear decrease in resistance with increasing temperature. They normally are encapsulated with a copper lug that is attached to the collector and storage tank. When connected to an electrical circuit of a differential controller, current flow varies in proportion to the difference between the resistances of the thermistors, resulting from the differences in liquid temperatures. The current change is amplified to close or open a relay and in turn operates the circulator-pump as necessary.

4.4.4 Pumps

Pumps-circulators typically used in solar DHW systems are of the centrifugal type as shown in Figure 4.12. This type of pump moves a fluid by sucking it into the center of a rapidly rotating disc (impeller) containing a series of blades. Creating a high velocity, the impeller imparts a centrifugal force that slings the liquid from the tips of the blades through the outlet. The frictional heat between the impeller and liquid is radiated out the pump body. These pumps are not self-priming, and as such, the liquid supply to the pump must be above its inlet. The materials used for the pump housing and wetted parts depends on whether the collector system is open or closed loop. Whereas the water-contacting parts of pumps used in closed loop systems may be made from iron, the parts used in open-loop systems must be manufactured from stainless steel or bronze. Failure to use a stainless steel or bronze body pump in an open-loop system with water will result in an unnecessarily short life for the pump because of corrosion.

4.4.5 Heat Exchangers

A heat exchanger is a device used to transfer heat from one medium to another without violating the integrity of either medium. Heat exchangers separate the heat transfer fluid in the collector loop from the domestic water in the storage tank. Heat transfer fluids should be nontoxic so as to require only a single-walled heat exchanger, depending on local plumbing codes. The coil in the storage tank is the most effective type of heat exchanger because it is located directly in the storage tank. The exchanger is normally a finned coil in the bottom of the tank. The rule of thumb for calculating maximum fin height is 65% of the tube diameter for most materials.

FIGURE 4.12 Centrifugal pump. Courtesy of Grundfos Pump Corporation. (For color version of this figure, the reader is referred to the online version of this book.)

The tank is coldest at its bottom and as heat from the transfer fluid is exchanged to the storage tank, the warmer water rises to the top of the tank by convection.

4.4.6 Storage Hot Water Tanks

To maximize the use of available insolation, it is necessary to have a method of energy storage that can retain and release solar derived energy on demand. Storage tanks typically are constructed of glass-lined steel, stone-lined steel, high-temperature fiberglass, or polyethylene, and they normally are available in standard sizes of 65, 80, 100, or 120 gallons. The size of storage needed,

Table 4.1 Example of Time Required to Heat Water Electrically

Tank Size (gallon)	Single Heating Element (W)	BTUs Consumed	Time to Heat (hr)
65	4500	54,145	3.5
80	4500	66,640	4.3
100	4500	83,300	5.4
120	4500	99,960	6.5

of course, depends on the amount of hot water consumed, as previously discussed.

Closed-loop solar DHW systems purchased as a package with collectors and storage tanks may be either provided with internal finned heat exchangers or a separate module containing an external heat exchanger and associated circulator-pump. In addition, storage tanks typically have a 4500 W electric element at the top of the tank that is used as a heating backup during periods of noninsolation.

It is interesting to note the amount of time it takes an electric water heater to completely heat a tank from 40–140 °F with one electric element of 4500 W. You can get an appreciation of the amount of electricity used by observing Table 4.1. This table is derived using the following assumptions:

$$\text{Temperature difference} \left(140 \,^\circ F - 40 \,^\circ F\right) = 100 \,^\circ F$$
$$\text{Specific Heat of Water} = 1 \text{ BTU/lb -} \,^\circ F$$
$$1 \, k\text{Wh} = 3414 \text{ BTUs}$$

To reduce heat loss, storage tank insulation should be a minimum of R-11. Increased values of insulation can be achieved by wrapping the tank with an additive insulative jacket from a commercially available kit. Table 4.2 illustrates storage tank heat loss in degrees Fahrenheit per hour and degrees Fahrenheit per day for the four common cylindrical tank sizes. As an example, for a 65-gallon tank, we will assume equal heat loss distribution on all surface areas, a tank temperature of 140 °F, an ambient air temperature of 50 °F, and a wall insulation of R-11. Heat loss Q of a storage tank can be expressed as follows:

$$Q = UA\Delta T$$

where:
$R = 11$ or $U = 1/R = 0.09$,
$A =$ total surface area of a cylinder (see Appendix A),
    Note: Appendix A has been included to provide additional helpful mathematical conversions and relationships, if needed.
$\Delta T =$ Temperature of water at 140 °F minus the ambient air at 50 °F.

Table 4.2 Storage Tank Heat Loss

Tank Size	Weight of Water (lb)	Typical Diameter (in)	Typical Height (in)	Total Surface Area (ft²)	Total Volume (ft³)	Heat Loss (BTUs/hr)	Temperature Loss (°F/hr)	Temperature Loss (°F/day)
65	541.5	24	60	37.7	15.7	305.4	0.56	13.5
80	666.4	26	60	41.4	18.4	335.3	0.50	12.0
100	833.0	26	69	46.5	21.2	376.7	0.45	10.8
120	999.6	28	69	50.7	24.6	410.7	0.41	9.8

Solving for Q (Heat Loss), we have the following:

$$Q = (0.09 \text{ BTU/ft}^2 - \text{hr} - {}^\circ\text{F}) \times (37.7 \text{ ft}^2) \times (90\,{}^\circ\text{F})$$

$$Q = 305.4 \text{ BTU/hr or } 7328.9 \text{ BTU/day (i.e.24 hr)}$$

Because 1 gallon of water weighs 8.33 lb, a 65-gallon tank would contain 541.5 lb of water. Because it takes one BTU/lb to raise the water temperature 1 °F, a loss of 541.5 BTUs will lower the tank 1 °F. Therefore, a 65-gallon tank will lose 13.5 °F [(7328.9 BTUs) ÷ (541.5 BTU/°F)] per day. In Table 4.2, as the volume-to-surface area ratio is increased, the total temperature drop is decreased.

4.4.7 Heat Transfer Fluids

In some instances, the type of solar DHW system chosen will determine the type of heat transfer fluid to be used. Other considerations include the geographic location of the system (protected from freezing or boiling), the potential of stagnation (fluid deterioration), component compatibility (some fluids will corrode or adversely affect aluminum, copper, and seal and gasket material), and environmental aspects (some fluids such as ethylene glycol are toxic and others have unpleasant odors when spilled).

The selection of the transfer fluid should not be a casual afterthought. These fluids are important to the life, design, and operation of the system. Each type of heat transfer fluid has its own unique properties, as illustrated in Table 3.2. Variations in viscosity, specific heat, coefficient of expansion, freezing point, boiling point, and flash point can determine the size and compatibility of many system components.

Ideally, the transfer fluid should have thermal stability and should not degrade at service temperatures or cause corrosive deterioration of system components. Corrosion can be caused by the composition of metals in the water (which varies from one locale to another), flow rate, the presence or lack of additives, or decomposition of the liquid at elevated temperature levels. Because of intermixed variables, the user, at a minimum, should monitor the pH periodically to ensure that the acidic content is low, preventing corrosion of the metal parts of the system. For copper and aluminum this pH factor should be >8. Let's discuss the types of transfer fluids listed in Chapter 3, Section 3.3, Table 3.2.

WATER
Untreated water is the least expensive and most readily available fluid. The four chemical compounds in water chemistry that cause problems in solar DHW systems include calcium carbonate, magnesium hydroxide, calcium silicate, and calcium sulfate. This is due to their decreased solubility with increasing temperatures. If freezing and boiling are not problems, if replacement water is readily at hand, and if there are no mineral hardness problems, water is an excellent transfer fluid. Because its specific heat is greater than any of the other fluids, it can deliver more BTUs to storage at a given flow rate. Whatever the local water

conditions, distilled or deionized water is recommended for drain-back systems with an in-tank heat exchanger. You should use a pump of bronze or stainless steel to prevent corrosion of the water-wetted surfaces.

GLYCOL–PROPYLENE

To overcome the deficiencies characteristic to water (such as freezing, boiling, and corrosion), propylene glycol can be added to water normally at a 50/50, 60/40, or 70/30 glycol to water ratio. This addition of glycols to water can solve many of the problems associated with water as a transfer fluid, by lowering the freezing temperature, raising the boiling point, and reducing corrosion as an inhibitor. Note, however, that glycol solutions can damage certain materials, such as butyl rubber membranes typically found in some types of expansion tanks. If stagnation occurs in the system, glycols can decompose rapidly at approximately 280 °F, forming sludge and organic acids. The higher the temperature, the more rapid the degradation. Glycol also breaks down in use. The buffers added to glycols are intended to prevent the pH from dropping and becoming acidic. Their effect is not permanent and if the glycol degradation is allowed to continue, the buffers will be depleted, and the solution in turn will become acidic. Because of this eventual acidic state, the solution should be monitored and a regular maintenance schedule should be maintained. The fluid should be changed at least once every 2 years based on pH results. It is the frequency of fluid change and service that increases the cost of using this transfer fluid over the lifetime of the system. An investment using an initially more costly synthetic hydrocarbon or silicone may be more economical because of less frequency of fluid change and service.

HYDROCARBONS

Hydrocarbon heat transfer fluids typically are categorized as either synthetic, paraffinic mineral oil, or aromatic refined mineral oil. Of these three types, only the synthetic is recommended.

Paraffinic mineral oils are petroleum-based heat transfer fluids, and their temperature range between boiling and freezing is greater than that of water. They are nonconducting and may have a higher viscosity than water. This type of transfer fluid is considered toxic and requires the use of a double-walled heat exchanger in a closed loop to preclude the danger of contaminating the potable water. Paraffinic mineral oils freeze at relatively high temperatures.

Aromatic mineral oils have lower viscosities than paraffins, allowing the use of smaller pumps. Because they have lower flash points, however, they are not as safe to use. Aromatics will dissolve roofing tar and most elastomer seals. Neoprene or Viton® seals should be used in pumps whenever paraffinic or aromatic hydrocarbons are used.

Synthetic hydrocarbons are not water miscible and therefore do not attack metals or elastomers as do aqueous solutions. They also do not develop excessive vapor pressure at normal operating temperatures as do water-based fluids. This type of heat transfer fluid normally will remain stable between 5 and

10 years, thereby decreasing maintenance. Like the aromatic mineral oils, if a synthetic hydrocarbon transfer fluid is spilled onto asphalt shingles or asphalt tile floors, it should be washed up quickly to prevent decomposition. Synthetic hydrocarbons typically are nontoxic.

SILICONES

Silicone heat transfer fluids are essentially inert, virtually nontoxic, will not freeze or boil, have no odor, and will not cause galvanic corrosion, and spills will not cause a degradation of roofing materials. These fluids also exhibit a high flash point. Although the initial cost of these fluids is higher than other transfer fluids, there is no need to monitor or replace the fluid periodically, and they have a life expectancy of 20 years or more.

There are, however, a few disadvantages to the fluid. Silicones have a lower heat capacity as illustrated in Table 3.2, as well as a higher viscosity, requiring twice the flow rate of most system fluids, and thus more pump horsepower. Silicones are incompatible with most expansion tanks fitted with neoprene or butyl rubber diaphragms. EPDM (Ethylene-Propylene-Diamine) rubber or Viton® materials should be used. Silicone fluids will readily leak through the smallest pipe joint soldering flaws, which normally would retain water or other fluids. Teflon tape, when used alone or with pipe dope, is unacceptable for threaded joints. All threaded connections must be sealed with either a Loctite® type pipe sealant or Dow Corning® fluorosilicone sealant to prevent leakage.

4.5 DETERMINING SOLAR ENERGY AVAILABILITY

Using the solar positions and insolation values for various latitudes from Table 4.3, we can easily determine the average surface daily insolation totals received at our particular latitude with the collector tilt angle equal to the latitude. For Billings, Montana, at a North Latitude of 46°, we would use the closest latitude from Table 4.3, which is 48°. The average of the surface daily totals in BTUs/ft²-day at 48° latitude (from the closest applicable values of 46° latitude and a 46° tilt for each month) in this case as developed from Table 4.3 is 1955 BTUs as shown in the following example.

Example from Table 4.3 Billings, Montana—48°Latitude (Collector Tilt Equal to Latitude) Surface Daily Totals in BTUs/ft²-day			
January	1478	August	2200
February	1972	September	2118
March	2228	October	1860
April	2266	November	1448
May	2234	December	1250
June	2204	Total annual amount	23,456
July	2200	Average surface daily total	1955°BTUs/ft²-day

Table 4.3 Solar Positions and Insolation Values for Various Latitudes

24 Degrees North Latitude

DATE	SOLAR TIME AM	PM	SOLAR POSITION ALT	AZM	BTUH/SQ. FT. TOTAL INSOLATION ON SURFACES NORMAL	HORIZ.	SOUTH FACING SURFACE ANGLE WITH HORIZ. 14	24	34	54	90
JAN 21	7	5	4.8	65.6	71	10	17	21	25	28	31
	8	4	16.9	58.3	239	83	110	126	137	145	127
	9	3	27.9	48.8	288	151	188	207	221	228	176
	10	2	37.2	36.1	308	204	246	268	282	287	207
	11	1	43.6	19.6	317	237	275	306	319	324	226
	12		46.0	0.0	320	249	296	319	332	336	232
	SURFACE DAILY TOTALS				2766	1622	1984	2174	2300	2360	1766
FEB 21	7	5	9.3	74.6	158	35	44	49	53	56	46
	8	4	22.3	62.2	263	116	135	145	151	150	102
	9	3	34.4	57.6	298	187	213	225	230	228	141
	10	2	45.1	44.2	314	241	273	286	291	287	168
	11	1	53.0	25.0	321	276	310	324	328	323	185
	12		56.0	0.0	324	288	323	337	341	335	191
	SURFACE DAILY TOTALS				3036	1998	2276	2396	2446	2420	1476
MAR 21	7	5	13.7	83.8	194	60	63	64	62	59	27
	8	4	27.2	76.8	267	141	150	152	149	142	64
	9	3	40.2	67.9	295	212	226	229	225	214	95
	10	2	52.3	54.8	305	266	285	288	283	270	120
	11	1	61.9	31.6	315	300	322	326	320	305	135
	12		66.0	0.0	317	312	334	338	333	317	140
	SURFACE DAILY TOTALS				3078	2270	2428	2456	2412	2298	1022
APR 21	6	6	4.7	100.6	40	7	5	3	3	2	2
	7	5	18.3	99.8	203	83	77	70	62	51	10
	8	4	32.0	89.0	256	160	157	149	137	122	16
	9	3	45.6	79.2	280	227	227	217	206	186	41
	10	2	59.0	71.8	292	275	275	259	259	237	61
	11	1	71.1	51.6	298	310	316	309	293	269	74
	12		77.6	0.0	299	321	328	321	305	280	79
	SURFACE DAILY TOTALS				3036	2454	2458	2374	2228	2016	488
MAY 21	6	6	8.0	108.4	86	22	14	10	9	7	5
	7	5	21.2	103.2	203	98	85	73	59	44	9
	8	4	34.6	98.5	248	171	159	145	127	106	15
	9	3	48.3	87.7	269	233	227	210	189	165	16
	10	2	62.0	76.9	280	280	275	261	240	211	22
	11	1	75.5		286	309	307	293	270	240	37
	12		86.0	0.0	288	319	317	304	281	250	57
	SURFACE DAILY TOTALS				3032	2556	2447	2286	2072	1800	246
JUN 21	6	6	9.3	111.6	97	29	19	13	12	11	13
	7	5	22.3	106.8	201	103	87	73	58	41	16
	8	4	35.5	102.6	242	173	158	142	122	99	18
	9	3	49.0	98.7	263	234	225	206	183	155	19
	10	2	62.6	95.0	274	280	273	253	229	199	22
	11	1	76.3	98.7	279	309	304	284	259	227	57
	12		89.4	0.0	281	313	310	294	265	236	204
	SURFACE DAILY TOTALS				3032	2626	2474	2266	1992	1700	204
JUL 21	6	6	8.2	109.0	81	23	16	11	10	9	6
	7	5	21.4	103.8	195	98	85	73	59	44	11
	8	4	34.8	99.2	239	169	157	143	125	104	18
	9	3	48.4	94.5	261	231	224	206	187	161	18
	10	2	62.1	89.0	272	278	270	256	235	206	21
	11	1	75.7		277	307	298	283	265	245	32
	12		86.6	0.0	280	317	312	302	275	245	36
	SURFACE DAILY TOTALS				2932	2526	2412	2250	2036	1766	246

DATE	SOLAR TIME AM	PM	SOLAR POSITION ALT	AZM	BTUH/SQ. FT. TOTAL INSOLATION ON SURFACES NORMAL	HORIZ.	SOUTH FACING SURFACE ANGLE WITH HORIZ. 14	24	34	54	90
AUG 21	6	6	5.0	101.3	35	7	5	4	3	2	2
	7	5	18.5	95.6	186	82	76	69	60	50	11
	8	4	32.2	89.7	241	158	154	146	134	118	16
	9	3	45.9	82.9	265	223	221	214	200	181	26
	10	2	59.3	73.0	278	273	268	261	295	261	42
	11	1	71.6	53.2	284	304	309	302	285	261	71
	12		78.3	0.0	286	315	320	313	296	272	75
	SURFACE DAILY TOTALS				2864	2408	2402	2316	2168	1958	470
SEP 21	7	5	13.7	83.8	173	57	60	60	60	56	55
	8	4	27.2	76.8	248	136	144	146	143	138	99
	9	3	40.2	67.9	278	205	218	221	217	206	143
	10	2	52.3	54.8	292	258	275	278	275	261	206
	11	1	61.9	31.6	299	291	311	315	309	295	261
	12		66.0	0.0	301	302	320	327	321	306	272
	SURFACE DAILY TOTALS				2878	2194	2342	2366	2322	2212	982
OCT 21	7	5	9.1	74.1	138	32	40	45	48	50	42
	8	4	22.0	66.7	247	111	129	139	144	145	99
	9	3	34.1	57.1	284	180	206	217	223	221	138
	10	2	44.7	43.8	301	234	265	277	282	279	165
	11	1	55.5	24.7	309	268	301	315	319	314	182
	12		55.5	0.0	311	279	314	328	332	327	188
	SURFACE DAILY TOTALS				2868	1928	2198	2314	2364	2346	1442
NOV 21	7	5	4.9	65.8	67	10	16	20	23	26	29
	8	4	17.0	58.4	232	82	108	123	135	142	124
	9	3	28.0	48.9	282	150	186	205	217	224	172
	10	2	37.3	36.3	303	203	244	265	278	283	204
	11	1	43.8	19.7	312	236	273	302	315	320	222
	12		46.2	0.0	315	247	293	315	328	332	228
	SURFACE DAILY TOTALS				2706	1610	1962	2146	2268	2324	1730
DEC 21	7	5	3.2	62.6	30	3	5	7	8	9	11
	8	4	14.9	55.3	225	71	99	116	129	137	130
	9	3	25.5	46.0	281	137	176	198	214	224	184
	10	2	34.3	33.7	304	189	234	258	275	283	217
	11	1	40.4	18.2	314	221	270	295	308	320	236
	12		42.6	0.0	317	232	282	308	325	332	243
	SURFACE DAILY TOTALS				2624	1474	1852	2058	2204	2286	1808

Sun

Tilt angle 0°   0°   90°

Tilt angle 45°   45°   0°   90°

Latitude ± 10°   Optimum tilt angle   0°   90°

Notes: a from ASHRAE Transactions, ground reflection not included

32 Degrees North Latitude

DATE	SOLAR TIME AM	PM	SOLAR POSITION ALT	AZM	NORMAL	HORIZ.	22	32	42	52	90
AUG 21	6	6	6.5	100.5	53	14	9	7	6	6	4
	7	5	19.1	92.8	190	85	77	69	60	50	12
	8	4	31.8	84.7	240	156	152	144	132	116	33
	9	3	44.3	75.0	263	216	220	212	197	178	65
	10	2	56.1	61.3	276	262	272	264	249	228	91
	11	1	66.0	38.4	282	292	305	298	281	257	107
	12		70.3	0.0	284	302	317	309	292	268	113
SURFACE DAILY TOTALS					2902	2252	2388	2296	2144	1934	736
SEP 21	7	5	12.7	81.9	163	51	56	55	52	52	50
	8	4	25.1	73.0	240	124	140	141	138	131	90
	9	3	36.8	62.1	272	188	215	215	211	201	114
	10	2	47.3	47.5	287	237	270	273	268	255	145
	11	1	55.0	26.8	294	268	306	309	303	289	164
	12		58.0	0.0	296	278	318	321	315	300	171
SURFACE DAILY TOTALS					2808	2014	2288	2308	2264	2154	1226
OCT 21	7	5	6.8	73.1	99	19	29	32	32	31	32
	8	4	18.7	64.0	229	90	120	128	133	134	104
	9	3	29.5	55.0	273	155	198	208	213	212	153
	10	2	38.7	39.1	293	204	257	269	273	270	188
	11	1	45.1	21.1	302	294	294	307	311	307	209
	12		47.5	0.0	306	247	306	320	324	318	217
SURFACE DAILY TOTALS					2696	1654	2100	2208	2252	2232	1588
NOV 21	7	5	1.5	65.4	2	0	0	0	1	1	1
	8	4	12.7	56.6	196	55	91	104	113	119	111
	9	3	22.6	46.1	263	118	173	195	202	208	176
	10	2	30.8	33.2	289	166	235	252	265	270	217
	11	1	36.2	17.6	301	192	270	291	303	307	241
	12		38.2	0.0	304	207	282	304	316	320	289
SURFACE DAILY TOTALS					2406	1280	1816	1980	2084	2130	1742
DEC 21	8	4	10.3	53.8	176	41	77	90	101	108	107
	9	3	19.8	43.6	257	102	161	180	195	204	183
	10	2	27.6	31.2	288	150	221	244	259	267	226
	11	1	32.7	16.4	301	180	258	282	298	305	251
	12		34.6	0.0	304	190	271	295	315	318	259
SURFACE DAILY TOTALS					2348	1136	1704	1888	2016	2086	1794

DATE	SOLAR TIME AM	PM	SOLAR POSITION ALT	AZM	NORMAL	HORIZ.	22	32	42	52	90
JAN 21	7	5	1.4	65.2	203	56	93	106	116	123	115
	8	4	12.5	56.5	269	118	175	193	203	217	181
	9	3	22.5	46.0	295	167	235	256	269	274	221
	10	2	30.6	33.1	306	198	273	295	308	317	245
	11	1	36.1	17.5	310	209	285	308	321	324	253
	12		38.0	0.0							
SURFACE DAILY TOTALS					2458	1288	1839	2008	2118	2166	1779
FEB 21	7	5	7.1	73.5	121	22	34	40	40	42	38
	8	4	19.0	64.4	247	95	127	136	141	141	108
	9	3	29.9	53.4	288	161	206	217	222	220	157
	10	2	39.1	39.4	306	212	266	278	283	279	193
	11	1	45.6	21.4	315	244	309	317	321	315	214
	12		48.0	0.0	317	255	316	330	334	328	222
SURFACE DAILY TOTALS					2872	1724	2188	2300	2345	2296	1644
MAR 21	7	5	12.7	81.9	185	54	60	60	59	56	32
	8	4	25.1	73.0	260	129	146	147	144	137	78
	9	3	36.8	62.1	290	194	222	224	220	209	119
	10	2	47.3	47.5	304	245	280	280	278	265	150
	11	1	55.0	26.8	311	277	317	321	315	300	170
	12		58.0	0.0	313	287	329	333	327	312	177
SURFACE DAILY TOTALS					3012	2084	2378	2403	2358	2246	1276
APR 21	6	6	6.1	99.9	66	14	10	7	6	5	3
	7	5	18.8	92.2	206	86	78	71	62	51	10
	8	4	31.5	84.0	255	158	156	147	136	120	35
	9	3	43.9	74.2	278	220	225	217	203	183	68
	10	2	55.7	60.3	290	267	280	272	256	234	95
	11	1	65.6	37.5	295	297	313	306	290	265	112
	12		69.6	0.0	297	307	325	318	301	276	118
SURFACE DAILY TOTALS					2990	2488	2556	2474	2296	2070	764
MAY 21	6	6	10.4	107.2	119	36	13	13	12	11	7
	7	5	22.8	100.1	211	107	88	75	60	44	12
	8	4	35.4	92.9	250	175	145	145	127	105	44
	9	3	48.1	84.7	269	233	223	209	188	163	105
	10	2	60.6	73.3	280	277	273	259	237	208	108
	11	1	72.0	51.9	285	305	305	290	268	237	99
	12		78.0	0.0	286	315	315	301	278	247	107
SURFACE DAILY TOTALS					3112	2582	2454	2284	2064	1788	469
JUN 21	6	6	12.2	110.2	131	45	15	16	15	14	9
	7	5	24.3	103.4	210	115	91	76	59	44	16
	8	4	36.9	96.8	245	180	159	143	122	99	19
	9	3	49.6	89.4	264	236	221	204	181	153	41
	10	2	62.2	79.7	274	279	268	251	227	197	56
	11	1	74.2	60.9	279	306	299	282	257	224	60
	12		81.5	0.0	280	315	310	292	267	234	74
SURFACE DAILY TOTALS					3084	2634	2436	2234	1990	1690	370
JUL 21	6	6	10.7	107.7	113	37	14	14	13	12	8
	7	5	23.1	100.6	203	107	87	75	60	44	14
	8	4	35.7	93.6	241	174	158	143	125	104	16
	9	3	48.4	85.5	261	231	220	205	185	159	51
	10	2	60.9	74.3	271	274	269	254	232	204	54
	11	1	72.4	53.3	277	302	300	285	262	224	69
	12		78.6	0.0	279	311	310	296	273	242	74
SURFACE DAILY TOTALS					3012	2558	2402	2250	2030	1754	458

Continued...

Table 4.3 Solar Positions and Insolation Values for Various Latitudes—continued

40 Degrees North Latitude

DATE	SOLAR TIME AM	SOLAR TIME PM	SOLAR POSITION ALT	SOLAR POSITION AZM	BTUH/SQ. FT. NORMAL	HORIZ	30	40	50	60	70	80	90 (WITH HORIZ.)
AUG 21	6	5	7.9	99.5	81	21	12	9	8	7			5
	7	4	19.3	90.0	191	87	74	63	60	49			12
	8	3	30.7	79.9	237	150	156	141	129	113			50
	9	2	41.8	67.9	260	205	216	207	193	175			89
	10	1	51.7	52.1	272	246	267	259	244	221			120
	11		59.3	29.7	278	273	300	292	276	252			140
	12		62.3	0.0	280	282	311	303	287	262			147
SURFACE DAILY TOTALS					2916	2354	2258	2104	1894				978
SEP 21	7	5	11.4	80.2	149	43	51	49	47	47			32
	8	4	22.5	69.6	230	109	133	131	124	114			84
	9	3	32.8	57.3	263	167	206	208	203	193			132
	10	2	41.6	41.9	280	211	262	265	260	247			168
	11	1	47.7	22.6	287	239	298	301	295	281			192
	12		50.0	0.0	290	249	310	313	307	292			200
SURFACE DAILY TOTALS					2708	1788	2210	2228	2182	2074			1416
OCT 21	7	5	4.5	72.3	48	14	17	17	17	16			16
	8	4	15.0	61.9	204	68	106	113	117	118			100
	9	3	24.5	49.8	257	126	185	195	200	198			160
	10	2	32.4	35.6	280	170	245	257	261	257			203
	11	1	37.6	18.7	291	199	283	295	299	294			229
	12		39.5	0.0	294	208	295	308	312	306			238
SURFACE DAILY TOTALS					2454	1348	2060	2160	2074				1654
NOV 21	8	4	8.2	55.4	136	28	63	72	78	82			81
	9	3	17.0	44.1	232	82	152	167	178	181			167
	10	2	24.0	31.0	268	126	215	233	245	249			219
	11	1	28.6	16.1	283	153	254	273	285	288			248
	12		30.2	0.0	288	163	267	287	298	301			258
SURFACE DAILY TOTALS					2128	942	1636	1778	1870	1908			1686
DEC 21	8	4	5.5	53.0	89	14	39	45	50	54			56
	9	3	14.0	41.9	217	65	135	152	164	171			163
	10	2	20.7	29.4	261	107	200	221	235	242			221
	11	1	25.0	15.2	280	134	239	262	276	283			252
	12		26.6	0.0	285	143	253	275	290	296			263
SURFACE DAILY TOTALS					1978	782	1490	1634	1740	1796			1646

DATE	SOLAR TIME AM	SOLAR TIME PM	SOLAR POSITION ALT	SOLAR POSITION AZM	BTUH/SQ. FT. NORMAL	HORIZ	30	40	50	60	70	80	90 (WITH HORIZ.)
JAN 21	8	4	8.1	55.3	142	28	65	74	81	85			84
	9	3	16.8	44.0	239	83	155	171	182	187			171
	10	2	23.8	30.9	274	127	218	237	249	254			223
	11	1	28.4	16.0	289	154	257	276	290	293			253
	12		30.0	0.0	294	164	270	291	303	306			263
SURFACE DAILY TOTALS					2182	948	1660	1810	1906	1944			1726
FEB 21	7	5	4.8	72.7	69	10	19	20	21	22			22
	8	4	15.4	62.2	224	73	114	122	127	127			107
	9	3	25.0	50.2	274	132	195	205	209	208			167
	10	2	32.8	35.9	295	178	256	267	271	267			210
	11	1	38.1	18.9	305	206	293	306	310	304			236
	12		40.0	0.0	308	216	306	319	323	317			245
SURFACE DAILY TOTALS					2640	1414	2060	2162	2202	2176			1730
MAR 21	7	5	11.4	80.2	171	46	55	51	51	51			35
	8	4	22.5	69.6	250	114	140	141	138	131			89
	9	3	32.8	57.3	282	173	215	219	213	207			138
	10	2	41.6	41.9	297	218	273	276	271	258			176
	11	1	47.7	22.6	305	247	310	313	307	293			200
	12		50.0	0.0	307	257	322	326	320	305			208
SURFACE DAILY TOTALS					2916	1852	2308	2330	2284	2174			1484
APR 21	6	6	7.4	98.9	89	20	11	8	7	7			4
	7	5	18.9	89.5	206	77	70	61	53	50			12
	8	4	30.3	79.3	252	152	153	133	117	117			53
	9	3	41.3	67.2	274	207	221	199	179	171			93
	10	2	51.2	51.4	286	250	275	252	229	258			126
	11	1	58.7	29.2	292	277	308	285	260	271			147
	12		61.6	0.0	293	287	320	296	271	296			154
SURFACE DAILY TOTALS					3092	2274	2412	2320	2168	1956			1022
MAY 21	5	7	1.9	114.7	1	0	0	0	0	0			0
	6	6	12.7	105.6	144	49	25	15	13	13			9
	7	5	24.0	96.6	216	89	89	76	65	60			13
	8	4	35.4	87.2	250	175	158	141	125	104			25
	9	3	46.8	76.0	267	227	221	206	186	160			60
	10	2	57.5	60.9	277	267	270	255	233	205			89
	11	1	66.2	37.1	283	293	301	287	264	234			108
	12		70.0	0.0	284	301	312	297	274	243			114
SURFACE DAILY TOTALS					3160	2552	2442	2264	2040	1760			724
JUN 21	5	7	4.2	117.3	22	4	2	1	1	1			1
	6	6	14.8	108.4	155	60	30	18	17	16			10
	7	5	26.0	99.7	216	123	92	77	59	57			14
	8	4	37.4	90.7	246	182	159	142	121	97			16
	9	3	48.8	80.2	263	233	202	179	151	121			47
	10	2	59.8	65.8	272	272	248	233	209	178			74
	11	1	69.2	41.9	277	296	278	262	238	206			92
	12		73.5	0.0	279	304	289	273	249	216			98
SURFACE DAILY TOTALS					3180	2648	2434	2224	1974	1670			610
JUL 21	5	7	2.3	115.2	2	0	0	0	0	0			0
	6	6	13.1	106.1	138	50	26	15	14	14			9
	7	5	24.3	97.2	208	114	89	75	60	44			14
	8	4	35.8	87.8	241	174	157	142	124	102			24
	9	3	47.2	76.7	259	225	218	203	182	157			58
	10	2	57.9	61.7	269	265	266	251	229	200			86
	11	1	66.7	37.9	275	290	296	281	258	228			104
	12		70.6	0.0	276	298	307	292	269	238			111
SURFACE DAILY TOTALS					3062	2534	2409	2230	2006	1728			702

48 Degrees North Latitude

DATE	AM	PM	SOLAR POSITION ALT	SOLAR POSITION AZM	BTUH/SQ. FT. NORMAL	HORIZ.	38	48	58	68	90
JAN 21	8	4	3.5	54.6	37	6	14	17	19	21	22
	9	3	11.0	42.6	185	83	120	132	140	145	139
	10	2	16.9	29.4	239	107	190	206	216	220	206
	11	1	20.7	15.1	261	115	231	249	260	263	243
		12	22.0	0.0	267	118	245	264	275	278	255
	SURFACE DAILY TOTALS				1710	596	1360	1478	1550	1578	1478
FEB 21	7	5	2.4	72.2	12	4	3	4	4	4	3
	8	4	11.6	60.5	188	49	95	102	105	106	96
	9	3	19.7	47.7	251	100	178	187	191	190	167
	10	2	26.2	33.3	278	139	240	251	255	251	217
	11	1	30.5	17.2	290	165	278	290	294	288	247
		12	32.0	0.0	293	173	291	304	307	301	258
	SURFACE DAILY TOTALS				2330	1080	1880	1972	2024	1978	1720
MAR 21	7	5	10.0	78.7	153	37	49	47	45	41	35
	8	4	19.5	66.8	236	96	131	132	129	122	96
	9	3	28.2	53.4	270	147	205	203	193	181	152
	10	2	35.4	37.8	287	187	266	263	252	234	195
	11	1	40.3	19.8	295	212	300	297	284	264	223
		12	42.0	0.0	298	220	312	309	296	275	232
	SURFACE DAILY TOTALS				2780	1578	2208	2228	2182	2074	1632
APR 21	6	6	8.6	97.8	108	27	13	8	8	7	4
	7	5	18.6	86.7	205	85	76	69	59	48	21
	8	4	28.5	74.9	247	142	149	141	129	113	69
	9	3	37.8	61.2	268	191	216	208	194	174	115
	10	2	45.8	44.6	280	228	268	260	245	223	152
	11	1	51.5	24.0	286	252	301	294	278	254	177
		12	53.6	0.0	288	260	313	305	289	264	185
	SURFACE DAILY TOTALS				3076	2106	2358	2266	2114	1902	1262
MAY 21	5	7	5.2	114.3	42	13	15	14	13	13	5
	6	6	14.7	103.7	162	61	27	16	15	13	13
	7	5	24.6	93.0	219	118	89	75	60	43	15
	8	4	34.7	81.6	248	171	156	140	123	101	45
	9	3	44.3	68.3	264	217	217	202	182	156	74
	10	2	53.0	51.3	274	252	265	251	229	200	105
	11	1	59.5	28.6	279	274	296	281	258	228	126
		12	62.0	0.0	281	281	306	292	269	238	133
	SURFACE DAILY TOTALS				3254	2482	2418	2234	2010	1728	982
JUN 21	5	7	7.9	116.5	77	21	22	15	14	15	11
	6	6	17.2	106.2	172	74	33	18	16	16	14
	7	5	27.0	95.8	220	129	95	75	58	39	15
	8	4	37.1	84.6	246	181	157	140	119	95	43
	9	3	46.9	71.6	261	225	216	199	175	147	83
	10	2	55.8	54.8	269	259	262	244	220	189	115
	11	1	62.7	31.2	274	278	291	273	248	216	137
		12	65.5	0.0	272	286	301	282	257	225	144
	SURFACE DAILY TOTALS				3156	2626	2420	2204	1944	1644	874
JUL 21	5	7	5.7	114.7	43	12	15	11	11	11	5
	6	6	15.2	104.1	156	62	28	18	14	14	13
	7	5	25.1	93.5	211	118	89	74	58	42	14
	8	4	35.1	82.1	240	171	154	138	121	99	43
	9	3	44.8	68.8	256	215	214	199	178	151	73
	10	2	53.5	51.9	266	250	261	246	224	195	105
	11	1	60.1	29.0	272	272	291	276	253	223	126
		12	62.6	0.0	275	279	301	286	263	232	133
	SURFACE DAILY TOTALS				3180	2536	2388	2200	1974	1694	955

DATE	AM	PM	SOLAR POSITION ALT	SOLAR POSITION AZM	BTUH/SQ. FT. NORMAL	HORIZ.	38	48	58	68	90
AUG 21	6	6	9.1	98.3	99	28	14	10	9	8	6
	7	5	19.1	87.2	190	85	75	67	58	47	20
	8	4	29.0	75.4	232	141	145	137	125	109	65
	9	3	38.4	61.8	254	189	210	201	187	168	110
	10	2	46.4	45.1	266	225	260	252	237	214	146
	11	1	52.2	24.3	272	248	285	293	268	244	169
		12	54.3	0.0	274	256	304	296	279	255	177
	SURFACE DAILY TOTALS				2558	2086	2300	2200	2046	1836	1208
SEP 21	7	5	10.0	78.7	131	35	44	43	40	35	31
	8	4	19.5	66.8	215	92	124	124	121	115	90
	9	3	28.2	53.4	251	142	196	197	193	183	143
	10	2	35.4	37.8	269	181	254	254	248	236	185
	11	1	40.3	19.8	278	205	287	289	284	269	212
		12	42.0	0.0	280	213	299	302	296	281	221
	SURFACE DAILY TOTALS				2528	1522	2102	2118	2070	1966	1546
OCT 21	7	5	2.0	71.9	4	1	1	1	1	1	1
	8	4	11.2	60.2	165	44	86	91	95	95	87
	9	3	19.3	47.4	233	94	167	176	180	178	157
	10	2	25.7	33.1	262	133	228	239	242	239	207
	11	1	30.0	17.1	274	157	266	277	281	276	237
		12	31.5	0.0	278	166	279	291	294	288	247
	SURFACE DAILY TOTALS				2154	1022	1774	1860	1890	1860	1626
NOV 21	8	4	3.6	54.7	36	5	17	19	21	22	22
	9	3	11.2	42.7	179	46	129	137	141	137	135
	10	2	17.1	29.5	233	83	186	202	212	215	201
	11	1	20.9	15.1	255	107	227	245	255	258	238
		12	22.2	0.0	261	115	241	259	270	272	250
	SURFACE DAILY TOTALS				1668	595	1335	1448	1518	1544	1442
DEC 21	9	3	8.0	40.9	140	27	98	109	117	122	121
	10	2	13.6	28.2	214	63	180	192	200	200	190
	11	1	17.3	14.4	242	86	207	226	235	241	231
		12	18.6	0.0	250	94	222	241	254	260	244
	SURFACE DAILY TOTALS				1944	446	1156	1250	1308	1364	1304

Example: Billings, Montana

Average surface daily totals at 48° tilt:

$$(1478+1972+2228+2266+2234+2204+2200+2200+2118+1860+1448+1250) \div 12 = 1955 \text{ BTU/ft}^2\text{-day}$$

Continued…

Table 4.3   Solar Positions and Insolation Values for Various Latitudes—continued

56 Degrees North Latitude

64 Degrees North Latitude

JUL 21 – DEC 21

DATE	AM	PM	ALT	AZM	NORMAL	HORIZ.	54	64	74	84	90
JUL 21	4	8	6.4	125.3	53	13	6	5	5	4	4
	5	7	12.1	112.4	128	44	32	13	11	10	9
	6	6	18.4	99.4	179	81	30	13	16	13	12
	7	5	25.0	86.0	211	118	86	77	56	38	28
	8	4	31.4	71.8	231	152	146	131	113	91	77
	9	3	37.3	56.3	245	182	245	186	166	141	124
	10	2	42.2	39.2	253	204	273	230	208	181	162
	11	1	45.4	20.2	257	218	282	258	236	207	187
	12		46.6	0.0	259	223	282	267	245	216	195
			SURFACE DAILY TOTALS		3372	2298	2280	2090	1864	1588	1400
AUG 21	5	7	4.6	108.8	29	6	3	3	2	2	2
	6	6	11.6	95.5	125	39	16	11	10	8	7
	7	5	17.6	81.9	181	77	69	61	52	42	35
	8	4	23.9	67.8	214	113	132	123	112	97	87
	9	3	29.6	52.6	234	144	190	182	169	150	138
	10	2	34.2	36.2	246	168	237	229	215	194	179
	11	1	37.2	18.5	252	183	268	260	244	222	205
	12		38.3	0.0	254	188	278	270	255	232	215
			SURFACE DAILY TOTALS		2808	1646	2108	1860	1662	1522	
SEP 21	7	5	6.5	76.5	77	16	25	25	24	23	21
	8	4	12.7	62.6	163	51	92	92	90	85	81
	9	3	18.1	48.1	206	83	159	159	156	147	141
	10	2	22.3	32.7	229	108	212	213	209	198	189
	11	1	25.1	16.6	240	124	246	248	243	230	220
	12		26.0	0.0	244	129	258	260	254	241	230
			SURFACE DAILY TOTALS		2074	892	1726	1736	1696	1608	1532
OCT 21	8	4	3.0	58.5	17	2	9	9	10	10	10
	9	3	8.1	44.6	122	26	86	91	93	92	90
	10	2	12.1	30.2	176	50	152	159	161	159	155
	11	1	14.6	15.2	201	65	193	201	203	200	195
	12		15.5	0.0	208	71	207	215	217	213	208
			SURFACE DAILY TOTALS		1238	358	1088	1136	1152	1134	1106
NOV 21	10	2	3.0	28.1	23	3	18	20	21	21	21
	11	1	5.4	14.2	79	12	70	76	79	80	79
	12		6.2	0.0	97	17	89	96	100	101	100
			SURFACE DAILY TOTALS		302	46	266	286	298	302	300
DEC 21	11	1	1.8	13.7	16	0	14	15	16	17	17
	12		2.6	0.0	24	2	20	22	24	24	24
			SURFACE DAILY TOTALS								

JAN 21 – JUN 21

DATE	AM	PM	ALT	AZM	NORMAL	HORIZ.	54	64	74	84	90
JAN 21	10	2	2.8	28.1	22	2	17	17	17	20	20
	11	1	5.2	14.1	81	12	72	77	80	81	81
	12		6.0	0.0	100	16	91	98	102	103	103
			SURFACE DAILY TOTALS		406	45	268	290	302	306	304
FEB 21	8	4	3.4	58.7	35	4	31	19	19	19	19
	9	3	8.6	44.8	147	31	103	108	111	110	107
	10	2	12.6	30.3	199	55	170	178	181	178	173
	11	1	15.1	15.3	222	71	212	220	223	219	213
	12		16.0	0.0	228	77	225	235	237	232	226
			SURFACE DAILY TOTALS		1452	400	1290	1286	1302	1282	1252
MAR 21	7	5	6.5	76.5	95	18	30	29	29	27	25
	8	4	12.7	62.6	185	54	101	102	99	94	89
	9	3	18.1	48.1	227	87	171	172	169	160	153
	10	2	22.3	32.7	249	112	227	229	224	213	205
	11	1	25.1	16.6	260	129	262	270	259	246	235
	12		26.0	0.0	263	134	274	277	271	258	246
			SURFACE DAILY TOTALS		2296	932	1286	1870	1830	1736	1556
APR 21	6	6	4.0	108.5	27	5	15	9	8	7	6
	7	5	10.4	95.1	133	37	70	63	54	43	37
	8	4	17.0	81.6	194	76	136	128	116	102	91
	9	3	23.3	67.5	228	112	197	189	175	158	145
	10	2	29.0	52.3	248	144	246	239	224	203	188
	11	1	33.5	36.0	260	169	278	270	255	233	216
	12		36.5	18.4	266	184	289	281	266	243	225
			97.6	0.0	268	190					
			SURFACE DAILY TOTALS		2296	1644	2176	2082	1936	1736	1594
MAY 21	5	7	5.8	125.1	51	11	13	11	10	9	8
	6	6	11.6	112.1	132	42	29	16	14	12	11
	7	5	17.9	99.1	185	79	86	72	56	39	28
	8	4	24.5?	85.7	218	117	148	133	115	94	80
	9	3	30.9	71.5	239	152	204	190	170	145	128
	10	2	36.8	56.1	252	182	249	235	213	186	167
	11	1	41.6	38.9	261	205	278	264	242	213	193
	12		44.9	20.1	265	219	288	267	251	222	201
			46.0	0.0	267	224					
			SURFACE DAILY TOTALS		3470	2236	2512	2124	1898	1624	1436
JUN 21	3	9	4.2	139.4			27	10	8	7	6
	4	8	9.0	126.4	93	27	60	16	13	11	10
	5	7	14.7	113.6	154	60	96	19	17	14	13
	6	6	21.0	100.8	194	96	34	72	55	36	23
	7	5	27.5	87.5	221	132	91	133	112	88	73
	8	4	34.0	73.3	239	166	150	187	164	137	119
	9	3	39.9	57.8	251	195	204	230	206	177	157
	10	2	44.9	40.4	258	217	247	258	233	202	181
	11	1	48.3	20.9	262	231	275	267	242	211	189
	12		49.5	0.0	263	235	288				
			SURFACE DAILY TOTALS		3650	2488	2342	2118	1862	1558	1356

Source: ©1974, ASHRAE (www.ASHRAE.org) used with permission from *ASHRAE Transactions, Vol. 80.*

4.6 SIZING A SOLAR DHW SYSTEM

At this point, we have determined the amount of energy we need to meet our hot water demands (explained in Chapter 3, Section 3.5.1). We also know the amount of radiant energy available per day (using Table 4.3), and we have an understanding of the efficiency of a solar DHW collector that can be provided by the solar installer's specification data. We therefore have enough information to calculate the number of solar DHW collectors required. If you wish to approximate your hot water requirements, instead of determining the precise amount of hot water used, you can simply use a rule-of-thumb method by assuming 20 gallons of water each for the first two persons and 15 gallons of water per person afterward. So a family of four would be using 70 gallons of water per day. Some dealers determine size by assuming one solar panel per person and 40 gallons of water per solar panel. Another easy approximation is to assume a collector efficiency, η, of 0.5 rather than reviewing the individual collector specification data sheets. So let's continue with our example at Billings, Montana.

The amount of energy delivered by a collector array can be obtained by the following algebraic equation:

$$\text{BTU's delivered from collector array} = (\eta) \times (A_c) \times (I)$$

where:
  η=the collector system efficiency (unitless),
  A_c=the effective collector area (ft^2), and
  I=the available solar radiation (BTU/ft^2-day).

From Chapter 3, Section 3.5.1, we determined that our typical family of four requires 55,395 BTUs to meet our domestic water needs for 1 day. This would be the amount of energy that should be delivered by the collector array. We also determined that the average solar insolation available, I, from Table 4.3 is 1955 BTUs/ft^2. We also will assume that the particular collector efficiency, η, specification indicates the panel efficiency is 0.5. We therefore can calculate the total area of the collectors required to produce sufficient energy.

$$\text{Therefore}: 55,395 \text{ BTUs} = (0.5) \times (A_c) \times \left(1955 \text{ BTUs}/\text{ft}^2\right)$$

$$\text{And} \quad A_c = \frac{55,395 \text{ BTUs}}{(0.5)\left(1955 \text{ BTU}/\text{ft}^2\right)}$$

$$\text{And} \quad A_c = 56.7 \text{ ft}^2$$

The minimum collector area (A_c) should be 56.7 ft^2. If the manufacturer's solar aperture or effective collector area is 22.2 ft^2 per collector, then we would require 2.6 collectors. This, of course, would round up to three collectors. An important thing to remember is to *not undersize* the system. If our results had indicated 2.9 or 3.0 collectors, then a four-panel system would be advisable to amply provide our hot water needs.

This method of determining the necessary collector area and thus the collector system output demonstrates that the system parameters of water storage capacity (i.e., amount of hot water required), water inlet temperature, collector panel efficiency, effective collector panel area, and available insolation are critical factors in determining collector array sizing.

The example of Table 4.4(a) and associated Appendix B worksheet of Table 4.4(b) will assist you in collecting and recording information for any particular location and in determining the proper collector sizing. Let's illustrate the use of these tables by determining the hot water consumption, the available solar contribution, and the proper collector sizing for our typical family of four in Billings, Montana. A line-by-line description follows for the example illustrated in Table 4.4(a) for the month of January.

Line A $= 31$ days/month

Line B $=$ Number of people in the household $= 4$

Line C $=$ Hot water used by 4 people $= (70 \text{ gallon/day}) \times (31 \text{ days/} \\ \text{month}) = 2170 \text{ gallon/month}$

Line D $=$ Hot water storage temperature $= 135\,°\text{F}$

Line E $=$ Inlet water temperature to storage is $40\,°\text{F}$ during January

Line F $=$ Line D $-$ Line E
$= (135\,°\text{F}) - (40\,°\text{F}) = 95\,°\text{F}$

Line G $= (8.33 \text{ lb/gallon}) \times (\text{line C}) \times (\text{line F})$
$= (8.33 \text{ lb/gallon}) \times (70 \text{ gal/day}) \times (95\,°\text{F})$
$= 55,395 \text{ BTUs/day}$

Line H $=$ Collector system efficiency (conservative average or directly from performance curve)
$= 0.6$

Line I $=$ Available insolation at $46°$. North latitude at optimum collector tilt of $46°$ (use Table 4.3, $48°$ north latitude at nearest optimum collector tilt of $48°$)
$= 1478 \text{ BTUs/ft}^2\text{-day}$

Line J $= (\text{line G}) - (\text{line H}) \times (\text{line I})$
$= (55,395 \text{ BTU/day}) \div (0.6) \times (1478 \text{ BTU/ft}^2\text{-day})$
$= 62.5 \text{ ft}^2$

This procedure can be followed to determine the minimum collector array size required to meet the domestic hot water demand for each month. In the example illustrated in Table 4.4(a), the yearly average collector area required is calculated to be 56.1 ft^2. If the effective collector aperture area is 22.2 ft^2, then three collectors should be installed, resulting in an actual total effective collector area installed of 66.6 ft^2. The total solar contribution and costs for heating domestic water for each month from such a collector array will be illustrated in

Table 4.4a Worksheet for Collector Sizing, Energy Consumption, and Solar Contribution

Worksheet for Collector Sizing, Energy Consumption, and Solar Contribution

Latitude __46°__
Collector Tilt Angle __48°__

Line	Evaluation Factors		Jan.	Febr.	Mar.	Apr.	May	June	Jul.	Aug.	Sept.	Oct.	Nov.	Dec.	Total Per Year
								Month							
A	Days in month		31	28	31	30	31	30	31	31	30	31	30	31	365
B	No. of people		4	4	4	4	4	4	4	4	4	4	4	4	4
1C	Hot water consumed (gallons; W_c)	Daily	70	70	70	70	70	70	70	70	70	70	70	70	
		Monthly	2170	1960	2170	2100	2170	2100	2170	2170	2100	2170	2100	2170	25,550
D	Storage temperature (T_s)		135	135	135	135	135	135	135	135	135	135	135	135	
E	Inlet water temperature (T_i)		40	40	40	45	45	50	50	50	50	45	40	40	
F	Avg. temp. increase D-E ($T_s - T_i$)		95	95	95	90	90	85	85	85	85	90	95	95	
G	BTU requirement (daily) 8.33* C*F		65395	65395	55595	52418	52418	60604	60604	60604	60604	52418	55595	65395	19,260,000
															Yearly average
2H	Collector system efficiency (n_s)		0.6	0.5	0.5	0.5	0.5	0.5	0.4	0.4	0.5	0.5	0.5	0.6	.5
3I	Available solar radiation (BTU/ft²-day)		1498	1972	2228	2266	2234	2204	2200	2200	2118	1860	1448	1250	1955
J	Array size (ft²) G ÷ (H*I)		62.5	56.2	49.7	46.3	47.0	45.0	56.3	56.3	46.8	56.4	76.5	73.9	56.1[4]

Actual total effective collector area installed (ft²) __66.6__

*Multiplication.
÷Division.

[1]Reference '20 + 20 + 15 + 15 + ...' supposition.
[2]Assume an average of 0.5.
[3]Reference Table 4–3
[4]Effective collector aperture area required.

Chapter 7. Software programs also are available on the Internet that can determine collector system sizing, eliminating the need to manually calculate your particular information as presented within this chapter. Unfortunately, links to these programs are not always available and are subject to change. Whether or not you use such online calculators, the preceding discussion will help you to understand the fundamentals involved in determining the amount of radiant energy received and the number and size of the solar panels needed for your particular hot water demand.

Chapter five

Solar Photovoltaic Systems

5.1 SOLAR PHOTOVOLTAIC FUNDAMENTALS

Simply stated, photovoltaic (PV) technology transforms sunlight into electricity. High-energy photons from the sun import energy to free electrons in semiconductor material thereby generating direct current (DC) electricity. A PV, or solar electric system, is made up of several PV solar cells. An individual PV cell is usually small, typically producing about 1 or 2 W of power. To boost the power output of PV cells, they are connected to form larger units called modules. Modules, in turn, can be connected to form even larger units called arrays, which can be interconnected to produce more power, and so on. Most solar PV modules up to 135 W are 12 V DC and many modules greater than 135 W are 21–40 V DC, designed mainly for grid-tie applications. PV modules are ganged up and mounted in series and in parallel depending upon the number of modules. A modest 1.2 kW (DC) system might employ twelve 100 W modules. A larger 6 kW system would need 60 such modules. On the other hand, a 6 kW system with 200 W modules would need only 30 modules, which means lower wiring and installation costs.

The most common PV array design uses flat-plate PV modules or panels. A typical flat-plate module design uses a substrate of metal, glass, or plastic to provide structural support in the back; an encapsulant material to protect the cells; and a transparent cover of a transparent polymer or glass. The majority of modules use a tempered "soda-lime" float glass similar to tempered window glass except that it has a much lower iron (Fe) content, allowing more transparency. The glass also can be treated in ways to change the index of refraction to minimize reflection. A sectional view of a typical flat-plate module is illustrated in Figure 5.1. The layers, in order from top to bottom, are cover film, solar cell, encapsulant, substrate, cover film, seal, gasket, and frame.

The panel arrays can be fixed in place or allowed to track the movement of the sun, and they respond to sunlight that is direct or diffuse, not unlike that of heat energy absorbed by solar hot water collectors discussed in Chapter 4, Section 4.3.1. Even with clear skies, the diffuse component of sunlight accounts for between 10% and 20% of the total solar radiation on a horizontal surface. On partly sunny days, up to 50% of that radiation is diffuse, and on cloudy days, up to 100%. The simplest PV array consists of flat-plate PV panels in a fixed

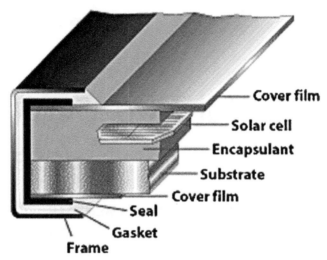

Cover film
Solar cell
Encapsulant
Substrate
Cover film
Seal
Gasket
Frame

FIGURE 5.1 Typical sectional view of a photovoltaic module. U.S. Department of Energy. (For color version of this figure, the reader is referred to the online version of this book.)

position oriented at a latitude tilt angle. The advantages of fixed arrays are that they lack moving parts that are subject to wear and failure, there is virtually no need for extra equipment, they are relatively lightweight, and they are less costly. A fixed-orientation angle to the sun therefore results in a more optimally cost-effective system for residential use.

5.2 BASIC PV SYSTEM COMPONENTS

A basic solar PV system consists of the following:

1. solar photovoltaic modules,
2. proper electrical disconnects and overcurrent protection systems, and
3. a string inverter or microinverters that change the DC generated electricity to alternating current (AC) used in most residences.

A system composed of these basic components with two arrays of modules, as illustrated in Figure 5.2, can generate electricity for your household and deliver any excess electricity back to the electrical grid for retail credit. Such an arrangement is called a grid-tied PV system. You should check with your utility company to ensure that you can connect a solar PV system to their electrical grid. Some rural electric cooperatives are exempt from the national law requiring interconnection. Ideally the utility company should buy back any excess electricity that your system produces at the same retail rate that you purchase from them. This arrangement is called "*net metering*," which provides a simple way to set up a grid-tied PV system. In this type of system, you normally have only one utility meter that is allowed to spin in either direction dependent on whether or not you are buying or selling energy. In a non non-net-metered system, the utility company normally will install a second utility meter to record

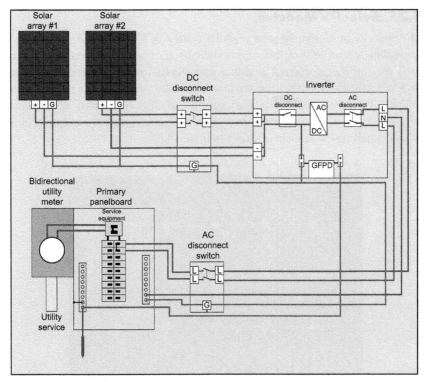

FIGURE 5.2 Grid-tied photovoltaic system with single inverter. Courtesy of www.Dream-HomeConsultants.com. (For color version of this figure, the reader is referred to the online version of this book.)

any excess energy that you sell back to them in which case you may be reimbursed energy costs only at a wholesale versus retail rate. Check with your particular state for the most current incentive programs and any limitations on the net-metered systems that can be connected to the grid in a specific utility service region. Such information normally is available from Internet database sites, such as the Database of State Incentives for Renewables and Efficiency (DSIRE) (www.dsireusa.org). A grid-tied type of system is designed to provide decades of economical and trouble-free electricity generated by the sun. Battery backup also can be incorporated in the system design; however, this will add to the complexity and cost of the system. Unless you are in a remote area without grid access, a battery backup system is not recommended for most residential applications. You should realize, however, that if the electrical power grid has an outage so will your PV system. The grid-tied solar electric inverter will shut down upon sensing a grid outage to prevent power from backfeeding to the power grid and injuring line workers. Because power outages are normally of short duration, it would be more cost-effective to install a backup generator than to include a more expensive battery backup system. Let's briefly discuss each of the basic component parts of a photovoltaic system.

5.2.1 Solar PV Modules

The majority of residential solar modules consist of PV cells made from either crystalline silicon cells or thin-film semiconductor material. Crystalline silicon cells are further categorized as either monocrystalline silicon cells that offer high efficiencies (13–19%) but are more difficult to manufacture or polycrystalline (also called multicrystalline) silicon cells that have lower efficiencies (9–14%) but are less expensive and easier to manufacture. An example of a monocrystalline PV module is shown in Figure 5.3.

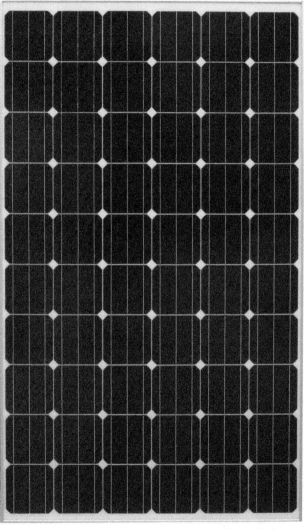

FIGURE 5.3 Suniva OPTimus 270 W monocyrstalline PV module (Model OPT 270-60-4-100). Photo courtesy of Suniva, Inc. (For color version of this figure, the reader is referred to the online version of this book.)

Thin-film solar cells, on the other hand, are manufactured by vaporizing and depositing thin layers of semiconductor material onto substrates, such as glass, ceramic, or metal. Although they absorb light more easily than crystalline silicon cells, they are much less energy production efficient (5–7%). They are, however, less costly to manufacture. The most efficient thin-film solar cells usually have several layers of semiconductor materials, such as gallium arsenide, that convert different wavelengths (i.e., colors) of light into electricity.

String ribbon manufactured modules also are available; however, current efficiencies are similar to thin-film modules requiring more surface area to produce the same output as the polycrystalline modules. Research advances in cell efficiency, materials, and methods of manufacturing continue to reduce costs and improve PV modules. The inherent inefficiency of PV modules is due to the fact that many of the electrons that have absorbed some energy from low-energy photons do not hold onto that energy long enough to absorb energy from another photon to free an electron. As a result, energy is lost as heat. To assist in cooling, these module arrays should be supported by framework that raises the entire system 3–6 in off the roof, allowing air to circulate keeping the system cool. An example of a 36 module array representing an 8.64 kW system is shown in Figure 5.4.

FIGURE 5.4 8.64 kW array composed of 36 Canadian Solar CSP6M monocrystalline silicon modules at 240 W each. Photo courtesy of ReVision Energy Corp. (For color version of this figure, the reader is referred to the online version of this book.)

5.2.2 Electrical Safety Disconnects

Electrical disconnects consist of additional switching that shuts off the AC power between the inverter and the grid, as well as a DC disconnect to safely interrupt the flow of electricity from the PV array to the inverter for system maintenance and troubleshooting possible system problems. An example of a PV disconnect system is shown in Figure 5.5. Utility companies that require these separate and overlapping circuit breakers want to ensure that the inverter drops offline during a power outage to prevent sending power to the grid, endangering repair personnel. These disconnects add costs and complexity to the photovoltaic system but ensure a redundancy to safety and overcurrent protection. Wiring should be of sufficient gauge (size) for the length of run to keep transmission losses to less than 3%. Normally, 12-gauge wire is sufficient for wiring between the solar array and a string inverter if less than 100 ft, and 10 gauge if more than 100 ft.

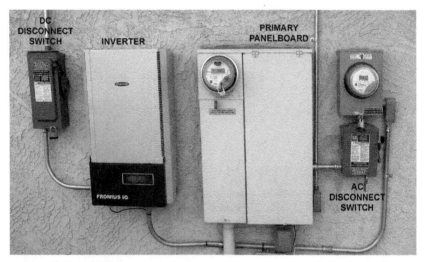

FIGURE 5.5 DC/AC Disconnect Arrangement. Courtesy of www.DreamHomeConsultants.com. (For color version of this figure, the reader is referred to the online version of this book.)

5.2.3 DC to AC inverters

A solar electric inverter is a component that converts DC electricity from the output of the PV array into grid-compliant AC electricity that is used in most homes. An inverter takes the DC power from the PV module array and causes it to oscillate until it matches the frequency of the power grid at 60 Hz (cycles per second). An inverter with ground fault protection also constantly checks for DC wiring shorts and bad connections, shutting the system down if problems are detected. If there is a power outage, the inverter will discontinue supplying electricity to the grid preventing electrical feedback to the power lines and personal injury to repair personnel. Most inverters have an efficiency of 85–96% depending on make and model. The power losses in the conversion of DC to AC as well as wire and switch-gear losses should be accounted for when determining the number of PV modules required. The inclusion of this loss factor when sizing a PV system is illustrated as step four of Table 5.3 (Section 5.4).

A traditional, centrally located inverter, as shown in Figure 5.6, is called a "string inverter" (because it is connected to a string of PV modules). It converts all of the DC current from the *entire* PV array into grid-compliant AC power. In a string inverter arrangement, the PV modules are connected in a series delivering accumulated DC voltage to the inverter for conversion into AC power that is fed into the power grid. The main shortcoming of a string inverter system is that every module in a typical string inversion system is limited by the weakest performing module. In other words, the maximum output performance of the string is defined by the poorest performing panel. For instance, if a single PV module is partially shaded and loses 40% of its output, every module in that string can become limited to the same 40% output. Inverters use a technique

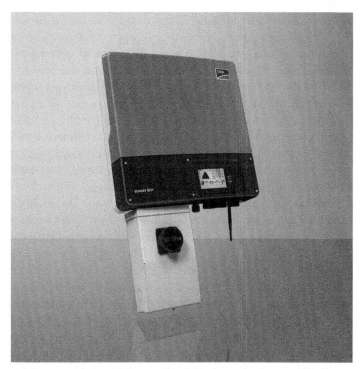

FIGURE 5.6 Sunny Boy 3000 TL-US/4000 TL-US/5000 TL-US string inverter. Photo courtesy of SMA Technology, AG. (For color version of this figure, the reader is referred to the online version of this book.)

known as maximum power point tracking (MPPT) to optimize PV output by adjusting applied loads. The PV array then can best use the available power at particular levels of available insolation. Because the effects of shading, snow covering, and module defects can cause variations in the output of an individual module, the inverter will change MPPT settings causing a divergence from an inverter's optimal performance. If a string inverter has multiple MMPT capabilities, as does the string inverter shown in Figure 5.6, the operating point with the highest performance can be found using more of the energy supply from the PV modules under shading-obstruction conditions.

Varying angles and nontraditional layouts and rooflines can present a problem for some string inverter systems because for those systems to function at their peak, all the PV modules need to have the same intensity of sunlight. All the modules, therefore, must be mounted at the same angle of incidence and facing the same direction. In addition, string inverters can have a limited selection of power ratings, which means that the power rating of the solar modules have to be matched with the power rating of the string inverter. This can place limitations on the option of expansion of the collector array.

An alternate type of inverter is a "**microinverter**" that converts the DC output of a *single* PV module into grid-compliant AC power. These are actually small inverters rated to handle the power output of a single panel. Each solar PV module has its own microinverter. Arrays of modules are connected in parallel with each other, and the AC power travels upstream through an ordinary branch circuit and then to the service panel. This type of microinverter system is a combination of multiple microinverters all along the branch circuits converting DC to AC power, all injecting their individual current supply. This individual parallel AC output structure as opposed to the DC series structure of a string inverter system has the advantage of isolating each panel. Reducing or losing the output from a single panel does not disproportionately affect the output from the entire array. Each microinverter is able to maintain optimum power by performing MPPT for its own individual module. The failure of a single panel or inverter in this type of system therefore will have minimal impact on overall system performance. PV module types and manufacturers can be mixed as long as they are compatible with the particular microinverter. The use of microinverters allows PV modules to be controlled independently, eliminating susceptibility to a reduction in system power output due to soiling, shading, and PV module defects. Unlike a single inverter functioning for an entire string of modules, there is no high-voltage wiring, and inverter outages only affect a small fraction of the PV system. A typical microinverter layout with rack mounts for the PV modules is shown in Figure 5.7. Using microinverters allows more flexibility in module arrays. Harsh weather conditions, however, are more likely to affect multiple electronic microinverters versus one string inverter. Costs also can be more prohibitive depending on the number of PV modules used, and selection of inverter configurations is primarily dependent on site conditions. Microinverter systems are scalable. If a project planner or builder wants to increase the capacity of a solar PV system at a later

FIGURE 5.7 Microinverters and mounting system for PV modules. Photo courtesy of ReVision Energy Corp. (For color version of this figure, the reader is referred to the online version of this book.)

point, additional modules can be added incrementally by simply extending the AC wiring to the next set of modules. Microinverter systems also can be monitored independently, making maintenance and upkeep simpler. A microinverter can be a good solution for installations with three or more roof orientations, difficult or rooftop shading issues and orientations, or very small systems under 3 kW.

The technology and design of inverters is continuously improving, and the use and cost of string inverters versus microinverters should be discussed with the solar energy installer or dealer for each particular application.

5.3 SOLAR PV COLLECTOR PERFORMANCE

Before we proceed with determining the number of PV modules required, we must review the performance output of the modules under consideration. Much like the SRCC (Solar Rating & Certification Corporation; defined in Section 4.4.2) rating for energy output from a solar hot water panel, we must understand the electrical rating used to compare PV modules. There are differences in how the modules are rated that must be considered to ensure consistencies for comparisons.

The nameplate ratings specified in the module's model number are derived by the manufacturer's factory testing protocol using **Standard Test Conditions** (STC), which keep the PV module test temperature at a constant 77 °F (25 °C).

This would work fine if the ambient air under actual working conditions were to remain fixed. That is not the case, however. On a hot summer day, the PV modules can heat up to more than 30° higher than the ambient air causing a reduction in the efficiency of the panel's power output. (PV electrical output decreases as temperature increases and vice versa.) Photocell voltage drops as temperature increases, so a PV module's output in real situations always will be lower than the power measured at the manufacturing facility where temperature is maintained at a constantly controlled 77 °F. The PV industry therefore developed a more real-life testing protocol called the "Photovoltaics for Utility Scale Applications" (PVUSA) Test Conditions or **PVUSA** Test Conditions (**PTC**) that ultimately were adopted by the **California Energy Commission (CEC)**. These more realistic testing conditions therefore are referred to as **CEC** ratings and are developed using testing conditions of 20 °C ambient temperature (68 °F), 10 m (32.81 ft) above ground level, with a wind speed of 1 m per second (3.281 ft/s). Refer to the **CEC** ratings (if available) when evaluating and comparing PV output power, rather than the STC ratings. This will provide a more conservative estimate. Table 5.1 illustrates the specification differences for several PV modules. This table is only a very small sampling, representing only a few of the hundreds of manufactured models available in a very competitive market.

Because the size of each model of PV module varies per manufacturer, you also need to consider the actual power generated per square foot. This information is important if you are limited on the area available for installation. Another factor to consider is the Temperature Coefficient of Power (**TcoP**), which is expressed as the percentage loss with each degree increase in temperature (Celsius). The closer this coefficient is to zero (i.e., the less negative the number), the better the hot weather performance. For instance, if a PV module has a TCoP of −0.30%, then the module will lose 6% of its output with a 20 °C increase in cell temperature (i.e., −0.30% per °C×20 °C=6%).

5.4 SIZING A SOLAR PV SYSTEM

Sizing a PV system is much easier to calculate than sizing a solar DHW system. Before you can determine the number of PV modules necessary to meet your energy requirements and size a system, you need to know the number of peak sun hours available in your area. Peak sun hours per day is defined as the equivalent number of hours per day when solar insolation averages 1 kW/m^2 (For example, five peak sun hours means that the energy received during total daylight hours equals the energy that would have been received had the radiation for 5 h been 1 kW/m^2) Table 5.2 represents the number of peak sun hours or radiant energy available in different states and cities that can be used to determine the number of PV modules needed. Do not confuse these numbers with the number of sunlight hours available from sunrise to sunset. Note that energy resource maps are available from websites, such as the National Renewable Energy Laboratory, previously known as the Solar Energy Research Institute.

Table 5.1 Examples of Photovoltaic Module Performance Ratings

Manufacturer	Cell Type	Model Number	Nameplate Rating (STC) (W)	PTC/CEC (W)	PTC/CEC (W/ft²)	Temperature Coefficient of Power (percent per degree Celsius)	Module Efficiency
Canadian solar	Monocrystalline (60 cells in series)	CS6P-240M	240	212	12.2 64.5 × 38.7 × 1.57 in (44.1 lbs)	−0.45	14.92%
REC	Poly-crystalline; 3 strings of 20 cells; 3 by-pass diodes	REC235PE	235	220 (est.)	12.4 (estimated) 17.74 ft² (39.6 lbs) 65.5 × 39.0 × 1.5 in	−0.46	14.2%
Samsung electronics	Monocrystalline (60 cells in series)	LPC250SM	250	225.9	13.8 64.17 × 36.6 × 1.86 in (41 lbs)	−0.48	15.62%
Sharp electronics Corp. (Sharp solar)	Poly-crystalline silicon (60 cells in series)	ND-224UC1	224	197.6	11.3 39 × 64.6 × 1.8 in (44.1 lbs)	−0.485	13.74%
Suniva	Mono-crystalline (60 cells in series)	OPT 270-60-4-100	270	240.3	13.8 17.46 ft² 65.04 × 38.66 in (39.5 lbs)	−0.420	16.6%
Sunpower Corp.	Mono-crystalline	SPR-X21-335-BLK	335	313.7	17.9 17.6 ft² 61.4 × 41.2 in (41 lbs)	−0.30	20.5%
Suntech power, Inc.	Poly-crystalline silicon (60 cells in series)	STP225-20/Wd	225	203.9	11.4 65.6 × 39.0 × 2.0 in (49.6 lbs)	−0.47	13.6%

Check manufacturer for latest and updated specifications.

Table 5.2 Peak Sun Hours per Day, National Averages.

State	City	Summer High	Winter Low	Year Round Average
AK	Fairbanks	5.87	2.12	3.99
AK	Matanuska	5.24	1.74	3.55
AL	Birmingham	–	–	5.2
AL	Montgomery	4.69	3.37	4.23
AR	Bethel	6.29	2.37	3.81
AR	Little Rock	5.29	3.88	4.69
AZ	Page	7.30	5.65	6.36
AZ	Phoenix	7.13	5.78	6.58
AZ	Tuscon	7.42	6.01	6.57
CA	Davis	6.09	3.31	5.10
CA	Fresno	6.19	3.42	5.38
CA	Inyokern	8.70	6.87	7.66
CA	La Jolla	5.24	4.29	4.77
CA	Los Angeles	6.14	5.03	5.62
CA	Riverside	6.35	5.35	5.87
CA	San Francisco	–	–	5.3
CA	Santa Maria	6.52	5.42	5.94
CA	Soda Springs	6.47	4.40	5.60
CO	Boulder	5.72	4.44	4.87
CO	Grandby	7.47	5.15	5.69
CO	Grand Junction	6.34	5.23	5.85
CO	Grand Lake	5.86	3.56	5.08
CT	Bridgeport	–	–	4.4
CT	Hartford	–	–	4.4
DC	Washington	4.69	3.37	4.23
DE	Wilmington	–	–	4.5
FL	Apalachicola	5.98	4.92	5.49
FL	Belle Island	5.31	4.58	4.99
FL	Gainsville	5.81	4.71	5.27
FL	Miami	6.26	5.05	5.62
FL	Tampa	6.16	5.26	5.67
GA	Atlanta	5.16	4.09	4.74

Table 5.2 Peak Sun Hours per Day, National Averages.—continued

State	City	Summer High	Winter Low	Year Round Average
GA	Griffin	5.41	4.26	4.99
HI	Honolulu	6.71	5.59	6.02
IA	Ames	4.80	3.73	4.40
ID	Boise	5.83	3.33	4.92
ID	Twin Falls	5.42	3.42	4.70
IL	Chicago	4.08	1.47	3.14
IL	Rockford	–	–	4.5
IN	Indianapolis	5.02	2.55	4.21
KS	Dodge City	4.14	5.28	5.79
KS	Manhattan	5.08	3.62	4.57
KS	Wichita	–	–	5.2
KY	Lexington	5.97	3.60	4.94
LA	Lake Charles	5.73	4.29	4.93
LA	Shreveport	4.99	3.87	4.63
MA	Blue Hill	4.38	3.33	4.05
MA	Boston	4.27	2.99	3.84
MA	E. Wareham	4.48	3.06	3.99
MA	Lynn	4.60	2.33	3.79
MA	Natick	4.62	3.09	4.10
MD	Silver Hill	4.71	3.84	4.47
ME	Caribou	5.62	2.57	4.19
ME	Portland	5.23	3.56	4.51
MI	Detroit	–	–	4.3
MI	E. Lansing	4.71	2.70	4.00
MI	Sault Ste. Marie	4.83	2.33	4.20
MN	Duluth	–	–	4.5C
MN	St. Cloud	5.43	3.53	4.53
MO	Columbia	5.50	3.97	4.73
MO	St. Louis	4.87	3.24	4.38
MS	Meridian	4.86	3.64	4.43
MT	Billings	–	–	5.0
MT	Glasgow	5.97	4.09	5.15

Continued...

Table 5.2 Peak Sun Hours per Day, National Averages.—continued

State	City	Summer High	Winter Low	Year Round Average
MT	Great Falls	5.70	3.66	4.93
MT	Summit	5.17	2.36	3.99
NC	Cape Hatteras	5.81	4.69	5.31
NC	Greensboro	5.05	4.00	4.71
ND	Bismark	5.48	3.97	5.01
ND	Fargo	–	–	4.6
NE	Lincoln	5.40	4.38	4.79
NE	N. Omaha	5.28	4.26	4.90
NH	Concord	–	–	4.6
NJ	Atlantic City	–	–	4.6
NJ	Newark	–	–	4.4
NJ	Sea Brook	4.76	3.20	4.21
NV	Ely	6.48	5.49	5.98
NV	Las Vegas	7.13	5.84	6.41
NY	Albany	–	–	4.3
NY	Binghamton	3.93	1.62	3.16
NY	Ithaca	4.57	2.29	3.79
NY	N.Y. City	4.97	3.03	4.08
NY	Rochester	4.22	1.58	3.31
NY	Schenectady	3.92	2.53	3.55
OH	Cleveland	4.79	2.69	3.94
OH	Columbus	5.26	2.66	4.15
OK	Oklahoma City	6.26	4.98	5.59
OK	Stillwater	5.52	4.22	4.99
OR	Astoria	4.76	1.99	3.72
OR	Corvallis	5.71	1.90	4.03
OR	Medford	5.84	2.02	4.51
PA	Harrisburg	–	–	4.6
PA	Pittsburg	4.19	1.45	3.28
PA	State College	4.44	2.79	3.91
RI	Newport	4.69	3.58	4.23
SC	Charleston	5.72	4.23	5.06

Table 5.2 Peak Sun Hours per Day, National Averages.—continued

State	City	Summer High	Winter Low	Year Round Average
SD	Rapid City	5.91	4.56	5.23
TN	Memphis	–	–	5.1
TN	Nashville	5.20	3.14	4.45
TN	Oak Ridge	5.06	3.22	4.37
TX	Austin	–	–	5.2
TX	Brownsville	5.49	4.42	4.92
TX	El Paso	7.42	5.87	6.72
TX	Fort Worth	6.00	4.80	5.43
TX	Midland	6.33	5.23	5.83
TX	San Antonio	5.88	4.65	5.30
UT	Flaming Gorge	6.63	5.48	5.83
UT	Salt Lake City	6.09	3.78	5.26
VA	Lynchburg	–	–	4.9
VA	Richmond	4.50	3.37	4.13
WA	Prosser	6.21	3.06	5.03
WA	Pullman	6.07	2.90	4.73
WA	Spokane	5.53	1.16	4.48
WI	Madison	4.85	3.28	4.29
WV	Charleston	4.12	2.47	3.65
WY	Cheyanne	–	–	5.3
WY	Lander	6.81	5.50	6.06

Source: Developed from the U.S. Department of Energy.

We now have the following information available to estimate the number of PV modules required:

1. We have reviewed the electrical utility bills and have determined how much electricity is used on an average daily basis as illustrated in Chapter 3, Section 3.5.2.
2. We know what the STC and CEC panel ratings mean (Section 5.3).
3. We can approximate the number of sun hours based on the closest city to us using Table 5.2, and
4. We understand there is a system performance loss that should be included with our calculations.

With the above information, you can now estimate the number of panels needed to supply 100% of your electricity requirements. Table 5.3 provides a step-by-step method for sizing a PV system. There are various other methods for determining the number of panels in an array, however, the example that follows provides an easy and practical method in which to calculate the number of Solar PV modules needed to generate electricity for your particular energy demands.

Table 5.3 Sizing Method for the Determination of the Number of Photovoltaic Modules

Sequence	Method	Example Calculation (i.e., Billings, Montana[1])	Example Result
Step 1	Annual kilowatt-hours from 12 months of utility bills	Annual kWh = 10,800 kWh	10,800 kWh/year (Annual demand)
Step 2	Average daily kWh	$\dfrac{10,800 \text{ kWh}}{\text{year}} \times \dfrac{1 \text{ year}}{365 \text{ days}}$	29.6 kWh/day (daily demand)
Step 3	Divide daily demand by peak Sun hours (Table 5.2)[1]	$29.6\dfrac{\text{kWh}}{\text{day}} \div {}^{1}5.0\dfrac{\text{h}}{\text{day}} = 5.92 \text{ kW}$	5.92 kW (System size solar output required to yield 100% of daily demand)
Step 4	Multiply step 3 by 1.15[2] to account for DC to AC inverter power and wire run losses (efficiencies)	5.92 kW × 1.15 = 6.81 kW	6.81 kW (System size output including system energy losses)
Step 5	Divide daily supplied solar energy system output (step 4) by the *CEC* wattage rating output per solar module	(Assume selection of a solar module with a PTC/CEC wattage output of 220 W where 6.81 kW = 6810 W) $6,810 \text{ W} \div 220\dfrac{\text{W}}{\text{panel}} = 30.95$	Total number of photovoltaic modules = 31 (number of modules required to produce 100% of electrical demand)

[1]Sun hours per day; National Average for Billings, MT (see Table 5.2).
[2]Multiply the solar output by 1.15 to adjust for efficiency losses to determine the number of modules required to produce 100% of the energy demand. The inverse is true if the number of modules is known due to limited roof area, in which case the known output for the array would be multiplied by .85 to determine the actual output of the array assuming efficiency losses of 15%.

An array positioned at the same angle as the latitude of the site will receive the maximum *average* annual solar radiation. If your electricity demand is significantly higher in the winter, then you can maximize your solar receipt by setting the array at latitude plus 15°. If your electricity demand is significantly more in the summer months, then set the array tilt angle at latitude minus 15°. The number of PV modules calculated using the method in Table 5.3 supplies 100% of the electrical demand. You can reduce the number of panels in the system to accommodate your particular budget and investment costs as discussed in Chapter 6. You also might be limited by the installation area available (e.g., dimensions of your roof).

5.5 ARRANGEMENT OF PV MODULES

PV modules are arranged somewhat like solar DHW collectors. Because most grid-tied string inverters have a minimum 200 V start-up requirement and PV modules typically generate between 28 V DC and 32 V DC, the panels are strung together in a series configuration. Think of the array like a series of car batteries. Each panel is wired such that the positive terminal of one panel is wired to the negative terminal of the next, which in turn is wired to the positive terminal of the next panel, and so on to each panel. This results in a voltage multiplier. For instance, if you wire two 12 V DC car batteries together in series, you end up with 24 V DC and the same current as that of a single battery. Because the inverters require a higher voltage, the panel array must be wired in series. This arrangement, however, can cause a reduction in performance if one module fails, cutting off power from modules downstream from that panel. As mentioned previously, shading can cause issues if one or more modules are not producing power, having a similar reduction effect on the total output of the string of panels.

Microinverters are used where shading conditions can occur on individual modules. Since the cost of a microinverter is less than the cost of a single string inverter, at 2013 prices, the cost of using 18 microinverters for 18 PV modules may be about the same as that of a single string inverter (something that should be considered for overall costs). Note that not all microinverters are compatible with all PV modules. Smart choices for compatibility must be made to determine which conditions and combinations would work best. One might question whether using microinverters might lead to increased electronic malfunctions, but some confidence can be restored with the normally longer warranty period for these devices.

5.6 ELECTRICAL INSTALLATION CONSIDERATIONS

Article 690 of the National Electrical Code specifically addresses the safety standards for installation of grid-tied PV systems. Adherence to those standards will reduce the hazards associated with such electrical installations and ensure the performance and longevity of the system. It is important to understand the dangers associated with high and low direct voltage and current produced by PV systems. To install a reliable electrical system, you should be familiar with electrical power systems codes and with DC currents and power systems. Just to mention a few concerns, you should understand issues with the following:

- Improper types of conductors
- Excessive voltage drop because of long runs or inadequate wire gauge sizing
- Unsafe wiring methods
- Inadequate placement of disconnects
- Improper system grounding
- Use of underrated components
- Improper use of AC components in DC applications (such as fuses and switches)

Ensure that the PV modules meet UL Standard 1703 and inverters meet UL Standard 1741. Some inverters may have their internal circuitry tied to their case and force the central grounding point to be at the inverter input terminals. This type of design may not be compatible with ground fault equipment and may not provide maximum surge protection. Each disconnect and overcurrent device as well as wiring insulation must have voltage ratings exceeding the system voltage rating. Unless you have an experienced electrical background, it is advisable to ensure proper installation is contracted by a certified North American Board of Certified Energy Practitioners installer with a Master's electrician's license.

Chapter six

Economic Criteria for Financial Decisions

This Chapter serves as a gateway between the technical and economic guidelines of this book. It connects the basic technical guidelines regarding energy relationships and solar DHW and PV systems presented in Chapters 2 through 5 with the economic guidelines and considerations relative to these alternative energy systems presented in Chapters 7 through 9. The use of Single Payment Compound Amount Factors, Single Payment Present Worth Factors, and Capital Recovery Factors are discussed with associated examples from actual system quotations, including cash and loan repayment scenarios. State and Federal tax credits, rebates, explanations of credits and deductions, equity loans, and the relationships between consumer price index and inflation are all factors that lead to a conceptual understanding of an economic analysis. Understanding the information in this Chapter will provide a supportive foundation in determining the actual payback, break-even costs, and savings associated with the systems discussed in Chapters 7, 8, and 9. Taking all these factors into consideration will determine actual cost savings by using solar DHW and PV systems versus conventional systems.

Before we analyze the investments and resulting payback of using solar as an alternative solution to either heating water or supplementing our electricity demands, let's discuss a few of the basic financial criteria involved that support making those determinations. This chapter explains the basic fundamental calculations involved so you can determine the actual cost savings of a solar energy alternative system versus a conventional fossil fuel—based energy system. The discrete rate-of-return equations in this chapter may look ominous at first, but they are only multiplication factors. By simply multiplying the appropriate numbers contained in Table 6.1 under the applicable interest rates and periods, we can calculate actual payback times for particular loans as well as determine present and future worth of an initial investment. To accurately assess actual savings and payback period by using solar energy to heat domestic water or to provide electricity, the "time value of money" should be considered. Understanding the present and future worth of money is important to realizing the constantly increasing rates of inflation for the cost of energy. The basic economic discussions in this chapter, therefore, provide a more comprehensive interpretation of independently addressing costs and payback for solar domestic hot water (DHW) and solar photovoltaics (PV).

Table 6.1 Discrete Rate-of-Return Factors

Years	Single Payment		Uniform Payment Series
(n)	SPCAF	SPPWF	CRF
0.5% (Interest Rate)			
1	1.005	0.9950	1.0050
2	1.010	0.9901	0.5038
3	1.015	0.9851	0.3367
4	1.020	0.9802	0.2531
5	1.025	0.9754	0.2030
6	1.030	0.9705	0.1696
7	1.036	0.9657	0.1457
8	1.041	0.9609	0.1278
9	1.046	0.9561	0.1139
10	1.051	0.9513	0.1028
11	1.056	0.9466	0.0937
12	1.062	0.9419	0.0861
13	1.067	0.9372	0.0796
14	1.072	0.9326	0.0741
15	1.078	0.9279	0.0694
16	1.083	0.9233	0.0652
17	1.088	0.9187	0.0615
18	1.094	0.9141	0.0582
19	1.099	0.9096	0.0553
20	1.105	0.9051	0.0527
21	1.110	0.9006	0.0503
22	1.116	0.8961	0.0481
23	1.122	0.8916	0.0461
24	1.127	0.8872	0.0443
25	1.133	0.8828	0.0427
26	1.138	0.8784	0.0411
27	1.144	0.8740	0.0397
28	1.150	0.8697	0.0384
29	1.156	0.8653	0.0371
30	1.161	0.8610	0.0360

Table 6.1 Discrete Rate-of-Return Factors—continued

Years	Single Payment		Uniform Payment Series
(*n*)	SPCAF	SPPWF	CRF
0.75% (Interest Rate)			
1	1.008	0.9926	1.0075
2	1.015	0.9852	0.5056
3	1.023	0.9778	0.3383
4	1.030	0.9706	0.2547
5	1.038	0.9633	0.2045
6	1.046	0.9562	0.1711
7	1.054	0.9490	0.1472
8	1.062	0.9420	0.1293
9	1.070	0.9350	0.1153
10	1.078	0.9280	0.1042
11	1.086	0.9211	0.0951
12	1.094	0.9142	0.0875
13	1.102	0.9074	0.0810
14	1.110	0.9007	0.0755
15	1.119	0.8940	0.0707
16	1.127	0.8873	0.0666
17	1.135	0.8807	0.0629
18	1.144	0.8742	0.0596
19	1.153	0.8676	0.0567
20	1.161	0.8612	0.0540
21	1.170	0.8548	0.0516
22	1.179	0.8484	0.0495
23	1.188	0.8421	0.0475
24	1.196	0.8358	0.0457
25	1.205	0.8296	0.0440
26	1.214	0.8234	0.0425
27	1.224	0.8173	0.0411
28	1.233	0.8112	0.0397
29	1.242	0.8052	0.0385
30	1.251	0.7992	0.0373

Continued...

Table 6.1 Discrete Rate-of-Return Factors—continued

Years	Single Payment		Uniform Payment Series
(n)	SPCAF	SPPWF	CRF
1% (Interest Rate)			
1	1.010	0.9901	1.0100
2	1.020	0.9803	0.5075
3	1.030	0.9706	0.3400
4	1.041	0.9610	0.2563
5	1.051	0.9515	0.20604
6	1.062	0.9420	0.1725
7	1.072	0.9327	0.1486
8	1.083	0.9235	0.1307
9	1.094	0.9143	0.1167
10	1.105	0.9053	0.10558
11	1.116	0.8963	0.0965
12	1.127	0.8874	0.0888
13	1.138	0.8787	0.0824
14	1.149	0.8700	0.0769
15	1.161	0.8613	0.07212
16	1.173	0.8528	0.0679
17	1.184	0.8444	0.0643
18	1.196	0.8360	0.0610
19	1.208	0.8277	0.0581
20	1.220	0.8195	0.05542
21	1.232	0.8114	0.0530
22	1.245	0.8034	0.0509
23	1.257	0.7954	0.0489
24	1.270	0.7876	0.0471
25	1.282	0.7798	0.04541
26	1.295	0.7720	0.0439
27	1.308	0.7644	0.0424
28	1.321	0.7568	0.0411
29	1.335	0.7493	0.0399
30	1.348	0.7419	0.03875

Table 6.1 Discrete Rate-of-Return Factors—continued

Years	Single Payment		Uniform Payment Series
(n)	SPCAF	SPPWF	CRF
1.25% (Interest Rate)			
1	1.013	0.9877	1.0125
2	1.025	0.9755	0.5094
3	1.038	0.9634	0.3417
4	1.051	0.9515	0.2579
5	1.064	0.9398	0.2076
6	1.077	0.9282	0.1740
7	1.091	0.9167	0.1501
8	1.104	0.9054	0.1321
9	1.118	0.8942	0.1182
10	1.132	0.8832	0.1070
11	1.146	0.8723	0.0979
12	1.161	0.8615	0.0903
13	1.175	0.8509	0.0838
14	1.190	0.8404	0.0783
15	1.205	0.8300	0.0735
16	1.220	0.8197	0.0693
17	1.235	0.8096	0.0657
18	1.251	0.7996	0.0624
19	1.266	0.7898	0.0595
20	1.282	0.7800	0.0568
21	1.298	0.7704	0.0544
22	1.314	0.7609	0.0523
23	1.331	0.7515	0.0503
24	1.347	0.7422	0.0485
25	1.364	0.7330	0.0468
26	1.381	0.7240	0.0453
27	1.399	0.7150	0.0439
28	1.416	0.7062	0.0425
29	1.434	0.6975	0.0413
30	1.452	0.6889	0.0402

Continued...

Table 6.1 Discrete Rate-of-Return Factors—continued

Years	Single Payment		Uniform Payment Series
(n)	SPCAF	SPPWF	CRF
1.5% (Interest Rate)			
1	1.015	0.9852	1.0150
2	1.030	0.9707	0.5113
3	1.046	0.9563	0.3434
4	1.061	0.9422	0.2594
5	1.077	0.9283	0.20909
6	1.093	0.9145	0.1755
7	1.110	0.9010	0.1516
8	1.126	0.8877	0.1336
9	1.143	0.8746	0.1196
10	1.161	0.8617	0.10843
11	1.178	0.8489	0.0993
12	1.196	0.8364	0.0917
13	1.214	0.8240	0.0852
14	1.232	0.8118	0.0797
15	1.250	0.7999	0.07494
16	1.269	0.7880	0.0708
17	1.288	0.7764	0.0671
18	1.307	0.7649	0.0638
19	1.327	0.7536	0.0609
20	1.347	0.7425	0.05825
21	1.367	0.7315	0.0559
22	1.388	0.7207	0.0537
23	1.408	0.7100	0.0517
24	1.430	0.6995	0.0499
25	1.451	0.6892	0.04826
26	1.473	0.6790	0.0467
27	1.495	0.6690	0.0453
28	1.517	0.6591	0.0440
29	1.540	0.6494	0.0428
30	1.563	0.6398	0.04164

Table 6.1 Discrete Rate-of-Return Factors—continued

Years	Single Payment		Uniform Payment Series
(n)	SPCAF	SPPWF	CRF
1.75% (Interest Rate)			
1	1.018	0.9828	1.0175
2	1.035	0.9659	0.5132
3	1.053	0.9493	0.3451
4	1.072	0.9330	0.2610
5	1.091	0.9169	0.2106
6	1.110	0.9011	0.1770
7	1.129	0.8856	0.1530
8	1.149	0.8704	0.1350
9	1.169	0.8554	0.1211
10	1.189	0.8407	0.1099
11	1.210	0.8263	0.1007
12	1.231	0.8121	0.0931
13	1.253	0.7981	0.0867
14	1.275	0.7844	0.0812
15	1.297	0.7709	0.0764
16	1.320	0.7576	0.0722
17	1.343	0.7446	0.0685
18	1.367	0.7318	0.0652
19	1.390	0.7192	0.0623
20	1.415	0.7068	0.0597
21	1.440	0.6947	0.0573
22	1.465	0.6827	0.0552
23	1.490	0.6710	0.0532
24	1.516	0.6594	0.0514
25	1.543	0.6481	0.0497
26	1.570	0.6369	0.0482
27	1.597	0.6260	0.0468
28	1.625	0.6152	0.0455
29	1.654	0.6046	0.0443
30	1.683	0.5942	0.0431

Continued...

Table 6.1 Discrete Rate-of-Return Factors—continued

Years	Single Payment		Uniform Payment Series
(n)	SPCAF	SPPWF	CRF
2% (Interest Rate)			
1	1.020	0.9804	1.0200
2	1.040	0.9612	0.5151
3	1.061	0.9423	0.3468
4	1.082	0.9238	0.2626
5	1.104	0.9057	0.21216
6	1.126	0.8880	0.1785
7	1.149	0.8706	0.1545
8	1.172	0.8535	0.1365
9	1.195	0.8368	0.1225
10	1.219	0.8203	0.11133
11	1.243	0.8043	0.1022
12	1.268	0.7885	0.0946
13	1.294	0.7730	0.0881
14	1.319	0.7579	0.0826
15	1.346	0.7430	0.07783
16	1.373	0.7284	0.0737
17	1.400	0.7142	0.0700
18	1.428	0.7002	0.0667
19	1.457	0.6864	0.0638
20	1.486	0.6730	0.06116
21	1.516	0.6598	0.0588
22	1.546	0.6468	0.0566
23	1.577	0.6342	0.0547
24	1.608	0.6217	0.0529
25	1.641	0.6095	0.05122
26	1.673	0.5976	0.0497
27	1.707	0.5859	0.0483
28	1.741	0.5744	0.0470
29	1.776	0.5631	0.0458
30	1.811	0.5521	0.04465

Table 6.1 Discrete Rate-of-Return Factors—continued

Years	Single Payment		Uniform Payment Series
(n)	SPCAF	SPPWF	CRF
2.5% (Interest Rate)			
1	1.025	0.9756	1.0250
2	1.051	0.9518	0.5188
3	1.077	0.9286	0.3501
4	1.104	0.9060	0.2658
5	1.131	0.8839	0.21525
6	1.160	0.8623	0.1816
7	1.189	0.8413	0.1575
8	1.218	0.8207	0.1395
9	1.249	0.8007	0.1255
10	1.280	0.7812	0.11426
11	1.312	0.7621	0.1051
12	1.345	0.7436	0.0975
13	1.379	0.7254	0.0910
14	1.413	0.7077	0.0855
15	1.448	0.6905	0.08077
16	1.485	0.6736	0.0766
17	1.522	0.6572	0.0729
18	1.560	0.6412	0.0697
19	1.599	0.6255	0.0668
20	1.639	0.6103	0.06415
21	1.680	0.5954	0.0618
22	1.722	0.5809	0.0596
23	1.765	0.5667	0.0577
24	1.809	0.5529	0.0559
25	1.854	0.5394	0.05428
26	1.900	0.5262	0.0528
27	1.948	0.5134	0.0514
28	1.996	0.5009	0.0501
29	2.046	0.4887	0.0489
30	2.098	0.4767	0.04778

Continued...

Table 6.1 Discrete Rate-of-Return Factors—continued

Years	Single Payment		Uniform Payment Series
(n)	SPCAF	SPPWF	CRF
3% (Interest Rate)			
1	1.030	0.9709	1.0300
2	1.061	0.9426	0.5226
3	1.093	0.9151	3535
4	1.126	0.8885	0.2690
5	1.159	0.8626	0.21835
6	1.194	0.8375	0.1846
7	1.230	8131	0.1605
8	1.267	0.7894	0.1425
9	1.305	0.7664	0.1284
10	1.344	0.7441	0.11723
11	1.384	0.7224	0.1081
12	1.426	0.7014	0.1005
13	1.469	0.6810	0.0940
14	1.513	0.6611	0.0885
15	1.558	0.6419	0.08377
16	1.605	0.6232	0.0796
17	1.653	0.6050	0.0760
18	1.702	0.5874	0.0727
19	1.754	0.5703	0.0698
20	1.806	0.5537	0.06722
21	1.860	0.5375	0.0649
22	1.916	0.5219	0.0627
23	1.974	0.5067	0.0608
24	2.033	0.4919	0.0590
25	2.094	0.4776	0.05743
26	2.157	0.4637	0.0559
27	2.221	0.4502	0.0546
28	2.288	0.4371	0.0533
29	2.357	0.4243	0.0521
30	2.427	0.4120	0.05102

Table 6.1 Discrete Rate-of-Return Factors—continued

Years	Single Payment		Uniform Payment Series
(n)	SPCAF	SPPWF	CRF
3.5% (Interest Rate)			
1	1.035	0.9662	1.0350
2	1.071	0.9335	0.5264
3	1.109	0.9019	0.3569
4	1.148	0.8714	0.2723
5	1.188	0.8420	0.22148
6	1.229	0.8135	0.1877
7	1.272	0.7860	0.1635
8	1.317	0.7594	0.1455
9	1.363	0.7337	0.1314
10	1.411	0.7089	0.12024
11	1.460	0.6849	0.1111
12	1.511	0.6618	0.1035
13	1.564	0.6394	0.0971
14	1.619	0.6178	0.0916
15	1.675	0.5969	0.08683
16	1.734	0.5767	0.0827
17	1.795	0.5572	0.0790
18	1.857	0.5384	0.0758
19	1.922	0.5202	0.0729
20	1.990	0.5026	0.07036
21	2.059	0.4856	0.0680
22	2.132	0.4692	0.0659
23	2.206	0.4533	0.0640
24	2.283	0.4380	0.0623
25	2.363	0.4231	0.06067
26	2.446	0.4088	0.0592
27	2.532	3950	0.0579
28	2.620	0.3817	0.0566
29	2.712	0.3687	0.0554
30	2.807	0.3563	0.05437

Continued...

Table 6.1 Discrete Rate-of-Return Factors—continued

Years	Single Payment		Uniform Payment Series
(n)	SPCAF	SPPWF	CRF
4% (Interest Rate)			
1	1.040	0.9615	1.0400
2	1.082	0.9246	0.5302
3	1.125	0.8890	0.3603
4	1.170	0.8548	0.2755
5	1.217	0.8219	0.22463
6	1.265	0.7903	0.1908
7	1.316	0.7599	0.1666
8	1.369	0.7307	0.1485
9	1.423	0.7026	0.1345
10	1.480	0.6756	0.12329
11	1.539	0.6496	0.1141
12	1.601	0.6246	0.1066
13	1.665	0.6006	0.1001
14	1.732	0.5775	0.0947
15	1.801	0.5553	0.08994
16	1.873	0.5339	0.0858
17	1.948	0.5134	0.0822
18	2.026	0.4936	0.0790
19	2.107	0.4746	0.0761
20	2.191	0.4564	0.07358
21	2.279	0.4388	0.0713
22	2.370	0.4220	0.0692
23	2.465	0.4057	0.0673
24	2.563	0.3901	0.0656
25	2.666	0.3751	0.06401
26	2.772	0.3607	0.0626
27	2.883	0.3468	0.0612
28	2.999	0.3335	0.0600
29	3.119	0.3207	0.0589
30	3.243	0.3083	0.05783

Table 6.1 Discrete Rate-of-Return Factors—continued

Years	Single Payment		Uniform Payment Series
(n)	SPCAF	SPPWF	CRF
4.5% (Interest Rate)			
1	1.045	0.9569	1.0450
2	1.092	0.9157	0.5340
3	1.141	0.8763	0.3638
4	1.193	0.8386	0.2787
5	1.246	0.8025	0.22779
6	1.302	0.7679	0.1939
7	1.361	0.7348	0.1697
8	1.422	0.7032	0.1516
9	1.486	0.6729	0.1376
10	1.553	0.6439	0.12638
11	1.623	0.6162	0.1172
12	1.696	0.5897	0.1097
13	1.772	0.5643	0.1033
14	1.852	0.5400	0.0978
15	1.935	0.5167	0.09311
16	2.022	0.4945	0.0890
17	2.113	0.4732	0.0854
18	2.208	0.4528	0.0822
19	2.308	0.4333	0.0794
20	2.412	0.4146	0.07688
21	2.520	0.3968	0.0746
22	2.634	0.3797	0.0725
23	2.752	0.3634	0.0707
24	2.876	0.3477	0.0690
25	3.005	0.3327	0.06744
26	3.141	0.3184	0.0660
27	3.282	0.3047	0.0647
28	3.430	0.2916	0.0635
29	3.584	0.2790	0.0624
30	3.745	0.2670	0.06139

Continued...

Table 6.1 Discrete Rate-of-Return Factors—continued

Years	Single Payment		Uniform Payment Series
(n)	SPCAF	SPPWF	CRF
5% (Interest Rate)			
1	1.0500	0.95238	1.0500
2	1.1025	0.90703	0.53780
3	1.1576	0.86384	0.36721
4	1.2155	0.82270	0.28201
5	1.2763	0.78353	0.23097
6	1.3401	0.74622	0.19702
7	1.4071	0.71068	0.17282
8	1.4775	0.67684	0.15472
9	1.5513	0.64461	0.14069
10	1.6289	0.61391	0.12950
11	1.7103	0.58468	0.12039
12	1.7959	0.55684	0.11283
13	1.8856	0.53032	0.10646
14	1.9799	0.50507	0.10102
15	2.0789	0.48102	0.09634
16	2.1829	0.45811	0.09227
17	2.2920	0.43630	0.08870
18	2.4066	0.41552	0.08555
19	2.5269	0.39573	0.08275
20	2.6533	0.37689	0.08024
21	2.7860	0.35894	0.07800
22	2.9253	0.34185	0.07597
23	3.0715	0.32557	0.07414
24	3.2251	0.31007	0.07247
25	3.3864	0.29530	0.07095
26	3.5557	0.28124	0.06956
27	3.7335	0.26785	0.06829
28	3.9201	0.25509	0.06712
29	4.1161	0.24295	0.06605
30	4.3219	0.23138	0.06505

Table 6.1 Discrete Rate-of-Return Factors—continued

Years	Single Payment		Uniform Payment Series
(n)	SPCAF	SPPWF	CRF
6% (Interest Rate)			
1	1.0600	0.94340	1.0600
2	1.1236	0.89000	0.54544
3	1.1910	0.83962	0.37411
4	1.2625	0.79209	0.28859
5	1.3382	0.74726	0.23740
6	1.4185	0.70496	0.20336
7	1.5036	0.66506	0.17914
8	1.5938	0.62741	0.16104
9	1.6895	0.59190	0.14702
10	1.7908	0.55839	0.13587
11	1.8983	0.52679	0.12679
12	2.0122	0.49697	0.11928
13	2.1329	0.46884	0.11296
14	2.2609	0.44230	0.10758
15	2.3966	0.41727	0.10296
16	2.5404	0.39365	0.09895
17	2.6928	0.37136	0.09545
18	2.8543	0.35034	0.09236
19	3.0256	0.33051	0.08962
20	3.2071	0.31180	0.08719
21	3.3996	0.29416	0.08501
22	3.6035	0.27751	0.08304
23	3.8197	0.26180	0.08128
24	4.0489	0.24698	0.07968
25	4.2919	0.23300	0.07823
26	4.5494	0.21981	0.07690
27	4.8223	0.20737	0.07570
28	5.1117	0.19563	0.07459
29	5.4184	0.18456	0.07358
30	5.7435	0.17411	0.07265

Continued...

Table 6.1 Discrete Rate-of-Return Factors—continued

Years	Single Payment		Uniform Payment Series
(n)	SPCAF	SPPWF	CRF
7% (Interest Rate)			
1	1.0700	0.93458	1.0700
2	1.1449	0.87344	0.55309
3	1.2250	0.81630	0.38105
4	1.3108	0.76290	0.29523
5	1.4026	0.71299	0.24389
6	1.5007	0.66634	0.20980
7	1.6058	0.62275	0.18555
8	1.7182	0.58201	0.16747
9	1.8385	0.54393	0.15349
10	1.9672	0.50835	0.14238
11	2.1049	0.46509	0.13336
12	2.2522	0.44401	0.12590
13	2.4198	0.41496	0.11965
14	2.5785	0.38782	0.11434
15	2.7590	0.36245	0.10979
16	2.9522	0.33873	0.10586
17	3.1588	0.31657	0.10243
18	3.3799	0.29586	0.09941
19	3.6165	0.27651	0.09675
20	3.8697	0.25842	0.09439
21	4.1406	0.24151	0.09229
22	4.4304	0.22571	0.09041
23	4.7405	0.21095	0.08871
24	5.0724	0.19715	0.08719
25	5.4274	0.18425	0.08581
26	5.8074	0.17220	0.08456
27	6.2139	0.16093	0.08343
28	6.6488	0.15040	0.08239
29	7.1143	0.14056	0.08145
30	7.6123	0.13137	0.08059

Table 6.1 Discrete Rate-of-Return Factors—continued

Years	Single Payment		Uniform Payment Series
(n)	SPCAF	SPPWF	CRF
8% (Interest Rate)			
1	1.0800	0.92593	1.0800
2	1.1664	0.85734	0.56077
3	1.2597	0.79383	0.38803
4	1.3605	0.73503	0.30192
5	1.4693	0.68058	0.25046
6	1.5869	0.63017	0.21632
7	1.7138	0.58349	0.19207
8	1.8509	0.54027	0.17401
9	1.9990	0.50025	0.16008
10	2.1589	0.46319	0.14903
11	2.3316	0.42888	0.14008
12	2.5182	0.39711	0.13270
13	2.7196	0.36770	0.12652
14	2.9372	0.34046	0.12130
15	3.1722	0.31524	0.11683
16	3.4259	0.29189	0.11298
17	3.7000	0.27027	0.10963
18	3.9960	0.25025	0.10670
19	4.3157	0.23171	0.10413
20	4.6610	0.21455	0.10185
21	5.0338	0.19866	0.09983
22	5.4365	0.18394	0.09803
23	5.8715	0.17032	0.09642
24	6.3412	0.15770	0.09498
25	6.8485	0.14602	0.09368
26	7.3964	0.13520	0.09251
27	7.9881	0.12519	0.09145
28	8.6271	0.11591	0.09049
29	9.3173	0.10733	0.08962
30	10.063	0.09938	0.08883

Continued...

Table 6.1 Discrete Rate-of-Return Factors—continued

Years	Single Payment		Uniform Payment Series
(n)	SPCAF	SPPWF	CRF
9% (Interest Rate)			
1	1.0900	0.91743	1.0900
2	1.1881	0.84168	0.56847
3	1.2950	0.77218	0.39505
4	1.4116	0.70843	0.30867
5	1.5386	0.64993	0.25709
6	1.6771	0.59627	0.22292
7	1.8280	0.54703	0.19869
8	1.9926	0.50187	0.18067
9	2.1719	0.46043	0.16680
10	2.3674	0.42241	0.15582
11	2.5804	0.38753	0.14695
12	2.8127	0.35553	0.13965
13	3.0658	0.32618	0.13357
14	3.3417	0.29925	0.12843
15	3.6425	0.27454	0.12406
16	3.9703	0.25187	0.12030
17	4.3276	0.23107	0.11705
18	4.7171	0.21199	0.11421
19	5.1417	0.19449	0.11173
20	5.6044	0.17843	0.10955
21	6.1088	0.16370	0.10762
22	6.6586	0.15018	0.10590
23	7.2579	0.13778	0.10438
24	7.9111	0.12640	0.10309
25	8.6231	0.11597	0.10181
26	9.3992	0.10639	0.10072
27	10.245	0.09761	0.09974
28	11.167	0.08955	0.09885
29	12.172	0.08216	0.09806
30	13.268	0.07537	0.09734

Table 6.1 Discrete Rate-of-Return Factors—continued

Years	Single Payment		Uniform Payment Series
(n)	SPCAF	SPPWF	CRF
10% (Interest Rate)			
1	1.1000	0.90909	1.1000
2	1.2100	0.82645	0.57619
3	1.3310	0.75131	0.40211
4	1.4641	0.68301	31,547
5	1.6105	0.62092	0.26380
6	1.7716	0.56447	0.22961
7	1.9487	0.51316	0.20541
8	2.1436	0.46651	0.18744
9	2.3579	0.42410	0.17364
10	2.5937	0.38554	0.16275
11	2.8531	0.35049	0.15396
12	3.1384	0.31863	0.14676
13	3.4523	0.28966	0.14078
14	3.7975	0.26333	0.13575
15	4.1772	0.23939	0.13147
16	4.5950	0.21763	0.12782
17	5.0545	0.19784	0.12466
18	5.5599	0.17986	0.12193
19	6.1159	0.16351	0.11955
20	6.7275	0.14864	0.11746
21	7.4003	0.13513	0.11562
22	8.1403	0.12285	0.11401
23	8.9543	0.11168	0.11257
24	9.8497	0.10153	0.11130
25	10.835	0.09230	0.11017
26	11.918	0.08391	0.10916
27	13.110	0.07628	0.10826
28	14.421	0.06934	0.10745
29	15.863	0.06304	0.10673
30	17.449	0.05731	0.10608

Note: SPCAF, Single Payment Compound Amount Factor; SPPWF, Single Payment Present Worth Factor; CRF, Capital Recovery Factor.

6.1 FUTURE WORTH OF MONEY

Let's illustrate this concept with an example. Consider a loan of $8000 (*P*), a present sum of money, to be paid back with one payment at the end of a 5-year period with an interest rate of 5% a year. The amount actually owed at the end of the first year is the original sum of $8000 plus the $400 interest cost for the use of capital, for a total of $8400. At the end of the second year, the amount owed is $8400 plus the 5% (for the use of capital in the amount of $420 of interest) for a total of $8820. This process of compounding continues as illustrated in Table 6.2 until the end of the 5-year loan period, at which time the original $8000 borrowed actually has cost $10,210.40, representing a combination of the principal and accrued interest.

Table 6.2 Cost of $8000 at 5% Compounded Interest

SPCAF from Table 6.1 (*P*) × (SPCAF)	Compounded Amount Due	Loan Term in Years
($8000) × (1.0500)	$8400.00	1
($8000) × (1.1025)	$8820.00	2
($8000) × (1.1576)	$9260.80	3
($8000) × (1.2155)	$9724.00	4
($8000) × (1.2763)	$10,210.40	5

The total interest paid of $2210.40 ($10,210.40−$8000.00) is the rate of return on the money loaned. The lender can say that the "future worth" of the $8000 loaned at 5% over 5 years is $10,210.40. The computation of Table 6.2 can be determined expeditiously using Eqn (6.1).

$$S = P(1 + i)^n \tag{6.1}$$

Where:
  S = a sum of money at a specified future date,
  P = a present sum of money,
  i = interest rate earned at the end of each period, and
  n = the number of interest periods.

The "time value of money" can be displayed graphically as in Figure 6.1. At the end of the first period of time, the time value of P is $P + Pi$ or $P(1+i)$; at the second interval of time, the time value of P is $P(1+i) + P(1+i)$ or $P(1+i)^2$. The sum S at the end of the nth period will result in Eqn (6.1). The factor $(1+i)^n$ is called the **single payment compound amount factor** (**SPCAF**).

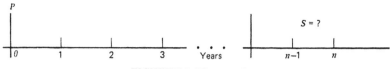

FIGURE 6.1 Time scales.

Equation 6.1 can then be rewritten as Eqn (6.2).

$$S = P\,(i - n\ SPCAF) \qquad (6.2)$$

We now can quickly calculate the compounded amount due over 5 years from Table 6.2 with one computation using the discrete rate-of-return factors from Table 6.1. For example, if \$8000 (*P*) is borrowed at 5% (*i*), over 5 years (*n*), the future worth (*S*), of the initial \$8000 can be found using Eqn (6.2) and the **SPCAF**, which can be obtained from Table 6.1 under the applicable interest rate, as follows:

$$S = P\,(i - n\ \text{SPCAF})$$

Where:

S = future worth of money,
P = \$8000 (the amount of money to be borrowed),
(*i* − *n* SPCAF) = (0.5 − 5 SPCAF) = 1.2763, and
S = \$8000 (1.2763) = \$10,210.40.

6.2 PRESENT WORTH OF MONEY

Because of inflation, future money is not as valuable as money at the present (especially if we keep printing paper money) and must be discounted by the factor $1/(1+i)^n$, which is called the **single payment present worth factor** (**SPPWF**). Simply stated, the present worth of money is the inverse of Eqn (6.1) and can be written as Eqn (6.3).

$$P = \frac{S}{(1+i)^n} \qquad (6.3)$$

Which can be rewritten as Eqn (6.4):

$$P = S\,(i - n\ SPPWF) \qquad (6.4)$$

For example, assuming a 5% inflation interest rate, the time value of a future sum of \$10,210 occurring 5 years from the initial investment can be found from Eqn (6.4) and the SPPWF from Table 6.1 under the applicable interest rate, as follows:

$$P = S\,(i - n\ SPPWF)$$

Where:

P=present worth of money,

S=$10,210 (future sum),

$(i-n\,\text{SPPWF})=(0.05-5\,\text{SPPWF})=0.78353$ (From Table 6.1), and

$P=(\$10,210)\times(0.78353)=\$8000.$

6.3 THE CAPITAL RECOVERY FACTOR

It is always convenient to discuss economics in terms of cash. What if a person does not have the cash, however, to purchase a solar DHW or PV system? Is it still cost-effective to finance a solar energy system? To evaluate an actual investment, we need to determine what the total cost of the system will be if the money is borrowed and yearly or monthly payments are made. The future series of end-of-period payments that will just recover a sum "P" over "n" periods with compound interest is illustrated in Figure 6.2. End-of-period payments can be determined using Eqn (6.5).

FIGURE 6.2 Capital recovery scale.

The factor by which a present capital sum "P" is multiplied to find the future repayment series "R" that will exactly recover it with interest is called the **capital recovery factor (CRF)**.

$$R=P\,(i-n\,\text{CRF}) \tag{6.5}$$

To illustrate the use of this factor, suppose $8000 ($P$) is borrowed for solar energy system equipment at 5% interest (i) compounded annually for a 5-year (n) term. The series of repayments for each year can be found using Eqn (6.5) and the CRF, which can be obtained from Table 6.1 under the applicable interest rate, as follows:

$$R=P\,(i-n\,\text{CRF})$$

Where:

R=repayment made at the end of each year,

P=$8000 (present sum),

$(i-n\,\text{CRF})=(0.05-5\,\text{CRF})=0.23097$, and

$R=(\$8000)\times(0.23097)=\$1847.76.$

This equates to a repayment of $153.98/month. Table 6.3 illustrates the cost of capital of $8000 at 5% interest with five end-of-year uniform payments for the recovery of capital.

Table 6.3 shows that the money on deposit at the beginning of each period (column 1) earns interest during that period (column 2), and the payment at the end of the period (column 4) repays the interest plus some of the principal (column 6). For example, the unpaid principal at the beginning of year 3 is $5032.09, the interest earned that year at 5% is $251.60, and the payment at the end of that year of $1847.76, consists of $251.60 in interest and $1596 (rounded into whole dollars) in principal.

Table 6.3 Visualizing the Capital Recovery Factor

Year	(1) Money Owed at Start of Year	(2) Interest Owed at End of Year	(3) (1)+(2) Principal and Interest Owed at End of Year	(4) Series of Repayments	(5) (3)−(4) Money Owed at End of Year after Repayment	(6) (4)−(2) Recovery Capital
1	$8000.00	$400.00	$8400.00	$1847.76	$6552.24	$1448
2	$6552.24	$327.61	$6879.85	$1847.76	$5032.09	$1520
3	$5032.09	$251.60	$5283.69	$1847.76	$3435.93	$1596
4	$3435.93	$171.80	$3607.73	$1847.76	$1759.97	$1676
5	$1759.97	$88.00	$1847.76	$1847.76	$0.00	$1760
					Total	$8000

The *actual cost of the solar energy system equipment* (in terms of present worth money at 5% interest compounded annually) is the sum of the initial amount of money borrowed plus the present worth of the interest due at the end of each repayment period. In this case, using Eqn (6.4) and the discrete rate-of-return factors from Table 6.1, the present worth of the money invested is shown in Table 6.4.

Table 6.4 Example: Actual Cost of a Solar Energy System – Borrowing $8000 at 5% Interest

(1) Number of Years	(2) Interest Owed at End of Year	(3) Single Payment Present Worth Factor (SPPWF)	(4) (2)×(3) Actual cost of a solar energy system
1	$400.00	0.95238	$380.95
2	$327.61	0.90703	$297.15
3	$251.60	0.86384	$217.34
4	$171.80	0.82270	$141.34
5	$88.00	0.78353	$68.95
Total present worth cost			$9105.73

Therefore, in a loan-repayment scenario of $8000 at 5% compounded interest with end-of-year repayments, the ***present worth of money*** invested for solar energy system equipment would be $9105.73 ($8000+$1105.73). This is the actual amount of money the solar energy system will cost you considering present worth costs at an interest rate of 5%. A solar energy system should be considered as an investment. As such, if money is borrowed to install such a system, the payments may be worked out with a bank so that repayments are approximately the same as the conventional monthly utility bill for either heating water or supplying electrical energy. Home equity loans or home equity lines of credit are two options to consider if financing a solar energy system.

6.4 SOLAR ENERGY TAX CREDITS

The Energy Tax Act enacted back in November of 1979 originally was passed by Congress as part of the National Energy Act. The objective of that law was to shift from oil and gas supply toward energy conservation and the use of renewable energy sources through taxes and tax credits. The law gave an income tax credit to private residents who adopted solar, wind, or geothermal sources of energy. The credit was equal to 30% of the cost of the equipment up to a maximum of $2000, as well as 20% of costs greater than $2000, up to a maximum of $10,000. Since that time, there have been many changes for allowable alternative tax incentives.

Through 2016, there are federal tax credit incentives in the United States allowing a 30% tax credit with no limit toward a rebate allowance of the cost of either a solar DHW or PV system. In addition, if the tax credit exceeds the amount of federal tax withheld, you could carry that amount over to the following tax year as a continued tax credit. So be sure to check the availability of such tax credits before purchasing a solar energy system to determine your true investment costs. Remember that similar tax credits were introduced back in the early 1980s and then ultimately removed. So it is not unprecedented that such federal incentives could be removed once again. The federal income tax credit can be obtained by filing Internal Revenue Service (IRS) Form 5695 with the yearly personal income tax IRS Form 1040. This credit is based on the cost of equipment and labor of a residential solar DHW or PV system.

Many states have introduced bills that continue to build momentum toward making solar energy more affordable for homeowners and businesses as well as create new jobs in the growing solar energy sector of the economy. Such laws include tax credits for the lease of solar equipment and power purchase agreements, statewide sales tax exemptions, income tax credits, and real property tax abatements for solar installations. Just like federal tax laws, however, such incentives are also subject to change. Instead of attempting to list the many variations in this book, it is recommended you browse the Internet for the most current information under topics, such as "state solar tax rebates" (e.g., www.dsireusa.org). It is important to understand that some state tax rebates may have installation requirements to be eligible for a tax rebate. Solar DHW installations may require the installer or dealer to have a master plumber's license, a master oil burner technician, or a propane and natural gas technician. PV installations may require the installer or dealer to have a master's electrician's license and be certified by the North American Board of Certified Energy Practitioners (NABCEP) or working with someone who is NABCEP certified.

State income taxes also are credited as a percentage return of personal income taxes on a solar investment depending on each state legislation. Savings on income taxes can be realized through the interest paid on borrowed money as a tax deduction from Schedule A of the IRS Form 1040. In some states, increased property taxes may result by adding a solar energy system if the assessed value of the property is increased. In other states, this additional value is exempt from property taxes. The availability of these tax credits and property-assessment exclusions should be investigated to ensure an accurate determination of actual system costs.

There always seems to be a bit of confusion about the difference between tax credits and tax deductions. It is important to recognize and understand this difference. First, a tax *credit* is *not* a tax *deduction*. *Deductions* are subtracted *from income* and represent only a percentage of an actual dollar reduction. *Credits* are subtracted *from taxes* owed and are a true dollar-for-dollar savings. For example, if you are entitled to a $1000 tax credit, and you owe $100 for income taxes, then you would subtract the total credit from the tax owed, and find you do not owe any taxes that year. In fact, you would carry over an additional $900 as a credit to be applied to the next year's taxes. Let's illustrate the difference of tax deductions and credit with another example. Suppose you earn $40,000/year and assume you have a personal earned tax liability of 20% of the income. The total tax liability therefore would be $8000. If you installed solar equipment in the amount of $6000, a 30% tax credit would reduce the $8000 tax liability by $1800. Figures 6.3(a) and 6.3(b) illustrate the tax credit received after completing IRS Form 5695 for a solar energy system cost of $6000. (Note that tax forms may change slightly from year to year.) You then would have a tax liability of $6200 versus $8000. If this $1800 credit was taken incorrectly as a tax deduction, the earned income would be reduced from $40,000 to $38,200, which would result in a tax liability of $7640 versus $6200. As you can see from this simple example, it is important to understand that a solar tax credit results in a tax credit, not a tax deduction, and is worth a lot more in your pocket.

Form **5695**	**Residential Energy Credits**	OMB No. 1545-0074
Department of the Treasury Internal Revenue Service	► See Instructions. ► Attach to Form 1040 or Form 1040NR.	**20**11 Attachment Sequence No. **158**

Name(s) shown on return	Your social security number
Resident Filing	x x x x x x x x x

Part I   Nonbusiness Energy Property Credit

1a Were the qualified energy efficiency improvements or residential energy property costs for your main home located in the United States? (see instructions) . . . . ►   **1a** ☑ Yes ☐ No

Caution: *If you checked the "No" box, you cannot claim the nonbusiness energy property credit. Do not complete Part I.*

b Print the complete address of the main home where you made the qualifying improvements.
Caution: *You can only have one main home at a time.*

1 Solar Lane
Number and street   Unit No.

Billings, MT 00000
City, State, and ZIP code

c Were any of these improvements related to the construction of this main home? . . . . . ►   **1c** ☐ Yes ☑ No

Caution: *If you checked the "Yes" box, you can only claim the nonbusiness energy property credit for qualifying improvements that were not related to the construction of the home. Do not include expenses related to the construction of your main home, even if the improvements were made after you moved into the home.*

2 Lifetime limitation. Amounts claimed in 2006, 2007, 2009, and 2010.

a	Amount, if any, from line 12 of your 2006 Form 5695	**2a**	0	
b	Amount, if any, from line 15 of your 2007 Form 5695	**2b**	0	
c	Amount, if any, from line 11 of your 2009 Form 5695	**2c**	0	
d	Amount, if any, from line 11 of your 2010 Form 5695	**2d**	0	
e	Add lines 2a through 2d. If $500 or more, **stop;** you cannot take the nonbusiness energy property credit			**2e**	0

3 Qualified energy efficiency improvements (original use must begin with you and the component must reasonably be expected to last for at least 5 years; do not include labor costs) (see instructions).

a	Insulation material or system specifically and primarily designed to reduce heat loss or gain of your home that meets the prescriptive criteria established by the 2009 IECC	**3a**	0
b	Exterior doors that meet or exceed the Energy Star program requirements	**3b**	0
c	Metal or asphalt roof that meets or exceeds the Energy Star program requirements and has appropriate pigmented coatings or cooling granules which are specifically and primarily designed to reduce the heat gain of your home	**3c**	0

d	Exterior windows and skylights that meet or exceed the Energy Star program requirements	**3d**	0		
e	Maximum amount of cost on which the credit can be figured	**3e**	$2,000		
f	If you claimed window expenses on your Form 5695 for 2006, 2007, 2009, or 2010, enter the amount from the Window Expense Worksheet (see instructions); otherwise enter -0-	**3f**	0		
g	Subtract line 3f from line 3e. If zero or less, enter -0-	**3g**	0		
h	Enter the smaller of line 3d or line 3g			**3h**	0
4	Add lines 3a, 3b, 3c, and 3h			**4**	0
5	Multiply line 4 by 10% (.10)			**5**	0

6 Residential energy property costs (must be placed in service by you; include labor costs for onsite preparation, assembly, and original installation) (see instructions).

a	Energy-efficient building property. Do not enter more than **$300**	**6a**	0
b	Qualified natural gas, propane, or oil furnace or hot water boiler. Do not enter more than **$150** . .	**6b**	0
c	Advanced main air circulating fan used in a natural gas, propane, or oil furnace. Do not enter more than **$50** .	**6c**	0
7	Add lines 6a through 6c	**7**	0
8	Add lines 5 and 7	**8**	0
9	Maximum credit amount. (If you jointly occupied the home, see instructions)	**9**	$500
10	Enter the amount, if any, from line 2e	**10**	0
11	Subtract line 10 from line 9. If zero or less, **stop;** you cannot take the nonbusiness energy property credit. .	**11**	500
12	Enter the smaller of line 8 or line 11	**12**	0
13	Limitation based on tax liability. Enter the amount from the Credit Limit Worksheet (see instructions) .	**13**	0
14	**Nonbusiness energy property credit.** Enter the smaller of line 12 or line 13. Also include this amount on Form 1040, line 52, or Form 1040NR, line 49	**14**	0

For Paperwork Reduction Act Notice, see your tax return instructions.   Cat. No. 13540P   Form **5695** (2011)

FIGURE 6.3 (a) Example of IRS Form 5695 for tax credit (sheet 1 of 2—tax year 2011)

6.5 HOME EQUITY LOANS AND LINES OF CREDIT

Not everyone has the cash available to finance a solar energy system. If you do have the cash available, it makes it that much easier to justify the payback. If you have to borrow the money, then the amount you have to pay in interest will subjugate and diminish the amount saved, but you will need to know to what extent. We will briefly discuss a couple of ways to finance a system, and then in more detail, we will examine the investment savings for both a solar DHW system (Chapter 7) and a PV system (Chapter 8). You should, of course, discuss

Form 5695 (2011)   Page **2**

Part II	**Residential Energy Efficient Property Credit** (See instructions before completing this part.)

Note. *Skip lines 15 through 25 if you only have a **credit carryforward from 2010.***

15	Qualified solar electric property costs	15	0
16	Qualified solar water heating property costs	16	6000
17	Qualified small wind energy property costs	17	0
18	Qualified geothermal heat pump property costs	18	0
19	Add lines 15 through 18	19	6000
20	Multiply line 19 by 30% (.30)	20	1800

21a Qualified fuel cell property. Was qualified fuel cell property installed on or in connection with your main home located in the United States? (See instructions) . . . . ▶ 21a ☐ Yes ☑ No

  Caution: *If you checked the "No" box, you cannot take a credit for qualified fuel cell property. Skip lines 21b through 25.*

  b Print the complete address of the main home where you installed the fuel cell property.

  _____   _____
  Number and street   Unit No.

  _____
  City, State, and ZIP code

22	Qualified fuel cell property costs	22	
23	Multiply line 22 by 30% (.30)	23	
24	Kilowatt capacity of property on line 22 above ▶ _____ x $1,000	24	
25	Enter the smaller of line 23 or line 24	25	
26	Credit carryforward from 2010. Enter the amount, if any, from your 2010 Form 5695, line 28 . .	26	0
27	Add lines 20, 25, and 26	27	1800
28	Enter the amount from Form 1040, line 46, or Form 1040NR, line 44 .	28	8000

29 **1040 filers:** Enter the total, if any, of your credits from Form 1040, lines 47 through 50; line 14 of this form; line 12 of the Line 11 Worksheet in Pub. 972 (see instructions); Form 8396, line 9; Form 8859, line 9; Form 8834, line 23; Form 8910, line 22; Form 8936, line 15; and Schedule R, line 22.

  1040NR filers: Enter the amount, if any, from Form 1040NR, lines 45 through 47; line 14 of this form; line 12 of the Line 11 Worksheet in Pub. 972 (see instructions); Form 8396, line 9; Form 8859, line 9; Form 8834, line 23; Form 8910, line 22; and Form 8936, line 15. | 29 | 0 |

30	Subtract line 29 from line 28. If zero or less, enter -0- here and on line 31	30	8000
31	**Residential energy efficient property credit.** Enter the smaller of line 27 or line 30. Also include this amount on Form 1040, line 52, or Form 1040NR, line 49	31	1800
32	Credit carryforward to 2012. If line 31 is less than line 27, subtract line 31 from line 27	32	

Form **5695** (2011)

FIGURE 6.3 (b) example of IRS Form 5695 for tax credit (sheet 2 of 2—tax year 2011).

direct loans with your local banks to determine the best approach. You will find that even if you borrow money to purchase the solar energy system, at today's low savings account interest rate at 0.5%, you actually will save more money by simply reducing your monthly energy expenditures by cost-effective alternative means.

A ***home equity line of credit*** allows you to draw funds from your bank up to a predetermined limit with an option to pay off as much of the line of credit used as you wish. A home equity line of credit normally carries a lower interest rate than a home equity loan, but its rate can fluctuate according to the prime rate. There are also normally no closing costs associated with establishing the line of credit, which is an additional cost factor to consider.

A *home equity loan*, on the other hand, provides you with a lump sum of money that has a fixed monthly payment over a predetermined period of time. You usually can choose between a variable interest rate of payment or a fixed rate, enabling you to budget a fixed monthly payment. For this type of loan, however, there are also closing costs to consider.

In both cases, the amount borrowed is based on such factors as the value of your home, your income, the remaining balance of your mortgage, and your credit history. It is the interest rates that makes these type of loans appealing because they are almost always lower than conventional bank loans because they are secured against the value of your home. In addition, the interest you pay on either type of loan is often tax deductible if you already meet the filing requirements of Schedule A and itemize deductions with your IRS Form 1040.

6.6 INFLATION AND CONSUMER PRICE INDEX

Inflation is a rising in the general level of prices of goods and services in an economy, or equivalently of a falling value of money. Inflation lessens the value of savings. As prices rise, the real value of purchasing power deteriorates. Savings accounts, annuities, and other fixed-value assets decline in real value. Inflation is not an easy thing to measure because prices for individual items do not rise evenly or proportionately. Various indexes have been devised to measure different aspects of inflation. The consumer price index (CPI) or the personal consumption expenditures index (PCE) tracks the prices consumers pay for things, and other indexes such as the producer price indexes (PPI) track prices the producers receive for goods and services they provide. The best measure of inflation for a given category depends on the intended use of the data. The CPI for all urban consumers is the most frequently reported statistic in the media and measures inflation as experienced by consumers in their daily living expenses for urban and metropolitan areas in the United States. It is therefore, normally, the better measurement for adjusting payments to consumer's current purchases and comparing them with those same purchases in an earlier period.

The Bureau of Labor Statistics (BLS) prices everything consumers purchase. Data collectors survey thousands of retail and service establishments to collect price data on thousands of items. All those prices then are categorized and weighted based on the average amount that the consumer spends on those categories. The percentage change from month to month is the rate of inflation expressed as a percentage. Comparing one period's price statistics against a previous month or previous year provides a crude measure of inflation (if the general level of prices has risen) or deflation (if the general level of prices has fallen). The overall inflation rate represents everything people spend money on, including things such as clothes, medical care, travel, haircuts, and food and energy. The food and energy categories are discarded, however, when calculating what is considered as the "core" inflation rate. Food and energy prices are excluded because they are historically highly volatile. The large changes in food and energy prices can occur because of supply disruptions, such as drought, and Organization of the Petroleum Exporting Countries cutbacks in oil production, respectively. An increase in the price of a single item, such an energy, therefore may cause a price index to rise. For this reason, many measures of core inflation have been developed from the basic price indexes, such as the CPI excluding food and energy and the CPI including only energy as shown in Table 6.5.

Table 6.5 Historical U.S. Inflation Rates from 2002 through 2012

Year	Yearly Average Consumer Price Index (CPI) for All Urban Consumers for All Items Less Food and Energy	Inflation Based on the CPI for All Urban Consumers for All Items Less Food and Energy[1]	Yearly Average CPI for All Urban Consumers for Energy	Inflation Based on the CPI for All Urban Consumers for Energy[2]
2002	190.44	2.32%	121.68	−5.77%
2003	193.23	1.46%	136.69	+12.33%
2004	196.64	1.77%	151.46	+10.80%
2005	200.87	2.15%	177.10	+16.93%
2006	205.92	2.51%	196.63	+11.03%
2007	210.73	2.34%	207.77	+5.66%
2008	215.57	2.30%	236.19	+13.68%
2009	219.24	1.70%	193.44	−18.10%
2010	221.34	0.96%	211.78	+9.48%
2011	225.01	1.66%	243.87	+15.16%
2012	229.75	2.11%	246.18	+0.95%

[1]Developed from Data Source: FRED, Federal Reserve Economics Data, Federal Reserve Bank of St. Louis; Consumer Price Index for All Urban Consumers: All Items Less Food and Energy (CPILFESL); U.S. Department of Labor; Bureau of Labor Statistics; http://research.stlouisfed.org/fred2/series/CPILFESL; accessed June 10, 2013.
[2]Developed from Data Source: FRED, Federal Reserve Economics Data, Federal Reserve Bank of St. Louis; Consumer Price Index for All Urban Consumers: Energy (CPIENGSL); U.S. Department of Labor; Bureau of Labor Statistics; http://research. stlouisfed.org/fred2/series/CPIENGSL?rid=10&soid=22; accessed June 10, 2013.

Calculating the inflation rate using the CPI is relatively simple. The BLS surveys thousands of prices all over the country every month to generate a CPI. For instance, the average CPI for energy was 136.69 in 2003 and 151.46 in 2004. To calculate the inflation factor in 2004, subtract the previous year's CPI (Y1) from the current year's CPI (Y2) and then divide that number by the previous year's CPI (Y1). The result is then multiplied by 100 to provide the result as a percentage. So we have the following equation:

$$\text{Inflation for current year } (2004) = \frac{Y2 - Y1}{Y1} \times (100)$$

Where:

    Y1 (previous year) = 136.69 (year 2003),
    Y2 (current year) = 151.46 (year 2004),

$$\text{And inflation for } 2004 = \frac{151.46 - 136.69}{136.69} \times (100) = 0.1080 \times (100) = 10.8\,\%$$

As an example of interpreting the effect of energy inflation rates shown in Table 6.5, the cost of energy in terms of 2013 prices is 72% higher than it was 10 years ago. If you add up the individual yearly inflation rates for energy shown in Table 6.5, you will find they equal an overall inflation rate of 72%. Therefore, over a 10-year period, the average energy inflation rate would be 7.2%/year. On the basis of this information and the continued volatility of rising energy prices, we will use a conservative energy inflation rate of 5% in the economic discussions of payback and break-even costs in Chapters 7, 8, and 9.

Chapter seven

The Economics of Solar Domestic Hot Water Systems

Let's determine whether the purchase of a solar hot water system is a cost-effective investment for your home. We previously calculated the amount of energy needed for your daily requirements as well as the amount of energy available at your location (Chapters 3 and 4). The most practical and cost-effective use of solar energy is its application to heating water. This chapter will discuss the cost savings and graphically illustrate examples of payback periods particular to heating water. If you are constructing a new residence, then you can consider using solar for hot water as well as radiant floor-space heating. In new construction, the total cost of a solar hot water system for both space heating and domestic hot water (DHW) applications would be less costly than attempting to retrofit your existing home. Because heating domestic water versus space heating has the most expedient return on investment, we will concentrate only on the DHW applications and discuss the financial advantage of such systems. Why should conventional methods of heating water continue to be used when there is a more economical method by using the sun's energy? You may not be able to have 100% of your hot water needs met by solar every day, but it can be a substantial supplement to your daily requirements. The technology of solar water heating is actually an old one; the present-day use of the technology continues to be new.

A solar DHW system simply must ultimately generate energy savings greater than its cost. That is a fairly easy concept to understand. Unless the solar energy system lowers the cost of living expenses, providing a better cash flow, it becomes a financial burden, rather than an investment. During the initial years, the system will have a negative impact on available monthly income from savings that otherwise would generate interest income or from interest paid on money borrowed to pay for the system. As years pass and energy prices rise, however, the system will produce an ever-increasing positive financial impact on *cash flow*. This is what makes a solar energy alternative system such an attractive investment. Solar DHW systems are initially more expensive than conventional hot water systems simply because more components are required. Installation labor costs are greater than conventional electric or gas water heaters or furnace-generated hot water system because a solar DHW system involves more facets of construction, including carpentry, plumbing, and electrical wiring. And although it just makes sense that solar energy system costs must be kept as low as possible for quicker payback time, you should not try to save money by purchasing inferior components. The system must be durable and have a life expectancy of at least 20 years to maintain good economics, and it *must not be undersized*. Price, quality, current and future energy costs, durability of components, and cost-saving tax credits, therefore, are among the many factors to consider.

As mentioned in Chapter 6, Section 6.4, some state tax rebates for solar DHW installations may require the installer or dealer to have a master plumber's license or to be a master oil burner technician or a propane and natural gas technician. The state rebate savings alone could determine whether it is worth your time and the needed skills to install your own system versus using a certified installer. Whether you install a system yourself or have it installed by a certified and licensed dealer, you will save money over a period of time by using solar as a supplementing energy source.

Although food and energy have been removed from the federal inflation equation (resulting in the actual deflation of the real inflation numbers), the costs of electricity and fossil fuels continue to escalate, placing a burden on household budgets already strained by real inflation. To determine the true "worth" of a solar DHW system, the estimated energy output of that system must first be calculated. We already calculated the amount of energy available at a specific location as well as the sizing of the collector array as illustrated previously using Chapter 4, Section 4.5, Table 4.3 and Section 4.6, Table 4.4(a). With that information, we can systematically determine the solar energy output by following the worksheet example guidelines of Table 7.1(a). (The associated worksheet of Table 7.1(b) is included in Appendix B to assist in recording information for determining solar energy output.) The information contained in Chapter 4, Table 4.4(a) can be used in conjunction with Table 7.1(a) to determine the British thermal unit (BTU) output. Because the sun does not shine at all locations 100% of the time, the monthly insolation available must be adjusted statistically by a percentage of possible sunshine as determined from the accumulation of historical data. Table 7.2 expresses the percentage of the maximum possible amount of sunshine reaching the earth's surface in the absence of clouds, fog, smoke, or other restrictions for selected locations. Note that possible sunshine percentages do not include the contribution of solar energy from diffuse radiation during cloudy days, which actually does provide additional energy gain.

This statistical percentage from the applicable City/State of Table 7.2 is entered in Table 7.1(a), line E. The remaining lines in the worksheet can be determined as annotated. Let's illustrate the use of Table 7.1(a) by continuing our discussion from Chapter 4 regarding a typical family of four in Billings, Montana. From the example of Table 4.4(a), the daily BTU requirement (line G), the daily available solar radiation (line I), the collector system efficiency

Table 7.1a Worksheet to Determine Solar Energy Contribution to DHW Energy Requirements

Latitude ___46°___
Collector Tilt Angle ___48°___
¹Effective Collector Aperture Area ___66.6___ ft²

Line	Evaluation Factors		Jan.	Feb.	Mar.	Apr.	May	Jun.	Jul.	Aug.	Sept.	Oct.	Nov.	Dec.	Total per Year
		(Month)	31	28	31	30	31	30	31	31	30	31	30	31	
A	Days in month		31	28	31	30	31	30	31	31	30	31	30	31	365
B	BTU requirement	Daily (BTU)	55395	55395	55395	52448	52448	49648	49648	49648	49648	52448	55395	55395	
		Monthly (MBTU)	1.72	1.55	1.72	1.57	1.63	1.49	1.54	1.54	1.49	1.63	1.66	1.72	19.26
C	Available solar radiation (BTU/ft²)	¹Daily	1478	1972	2228	2246	2234	2204	2200	2200	2118	1860	1448	1250	
		Monthly	45818	55216	69083	67380	69254	66120	68200	68200	63540	57660	44640	38750	
D	Collector array output (MBTU) (Collector area), °C		3.05	3.68	4.60	4.53	4.61	4.40	4.54	4.54	4.23	3.84	2.89	2.58	
E	Mean percentage of possible sun (Table 7.2)		0.47	0.53	0.62	0.60	0.61	0.64	0.77	0.75	0.68	0.62	0.46	0.44	0.62
F	Collector array output E * D (MBTU's)		1.43	1.95	2.85	2.72	2.81	2.82	3.50	3.41	2.88	2.38	1.33	1.14	29.22.2
G	¹Collector system efficiency (Ns)		0.6	0.5	0.5	0.5	0.5	0.5	0.4	0.4	0.5	0.5	0.5	0.6	0.5
H	Estimated system output, G * F (MBTU)		0.86	0.98	1.43	1.36	1.41	1.41	1.40	1.36	1.44	1.19	0.67	0.68	14.19
I	Percentage (%) solar contribution, H ÷ B		50	63	83	87	87	95	91	88	97	73	40	40	74.5

¹From Table 4.4a
* Multiplication.
÷ Division.

Table 7.2 Mean Percentage (%) of Possible Sun, through 2009

City/State	Years	Jan	Feb	Mar	Apr	May	Jun	Jul	Aug	Sep	Oct	Nov	Dec	Annual
Birmingham, AL	34	42	50	54	63	66	65	59	63	61	66	55	46	58
Montgomery, AL	45	47	52	59	65	63	62	61	63	62	64	55	49	59
Anchorage, AK	40	34	42	50	50	50	46	43	39	38	36	32	27	41
Juneau, AK	33	32	32	37	39	39	34	31	32	26	19	23	20	30
Nome, AK	40	40	55	54	54	50	43	37	32	36	34	31	34	42
Fort Smith, AR	51	50	55	56	60	62	69	73	72	66	65	54	50	61
Little Rock, AR	32	46	54	57	62	68	73	71	73	68	69	56	48	62
North Little Rock, AR	31	66	67	74	78	76	81	82	80	81	76	67	63	74
Flagstaff, AZ	15	77	73	76	82	88	86	75	76	81	79	75	73	78
Phoenix, AZ	101	78	80	84	89	93	94	85	85	89	88	83	77	85
Tucson, AZ	53	80	82	86	90	92	93	78	80	87	88	84	79	85
Yuma, AZ	42	84	87	90	94	95	97	91	91	93	92	87	82	90
Eureka, CA	84	43	46	52	57	58	59	55	51	55	50	44	41	51
Fresno, CA	46	47	65	77	85	90	95	97	96	94	88	66	46	79
Los Angeles, CA	32	69	72	73	70	66	65	82	83	79	73	74	71	73
Redding, CA	10	72	82	85	90	91	94	97	97	94	92	84	73	88
Sacramento, CA	46	48	65	74	82	90	94	97	96	93	86	66	49	78

San Diego, CA	56	72	71	70	68	59	58	68	70	69	68	75	73	68
San Francisco, CA	38	56	62	69	73	72	73	66	65	72	70	62	53	66
Denver, CO	49	71	69	69	67	64	70	71	71	74	72	64	67	69
Grand Junction, CO	56	61	65	65	70	73	81	79	77	79	74	63	61	71
Pueblo, CO	61	75	74	74	74	73	78	79	78	80	79	72	71	76
Hartford, CT	42	53	56	57	55	57	60	62	62	59	57	45	47	56
Washington, D.C.	50	46	50	55	57	58	64	62	62	61	59	51	46	56
Apalachicola, FL	57	58	61	65	74	77	71	64	64	66	74	67	57	67
Jacksonville, FL	50	58	62	68	73	70	66	65	64	58	60	60	54	63
Key West, FL	38	74	77	82	84	82	76	77	76	72	71	71	70	76
Miami, FL	20	66	68	74	76	72	68	72	71	70	70	67	63	70
Pensacola, FL	5	48	53	61	63	67	67	57	58	60	71	64	49	60
Tampa, FL	57	65	66	71	75	75	67	62	62	62	66	64	61	66
Atlanta, GA	65	49	54	58	66	68	67	63	64	62	66	58	50	60
Macon, GA	48	56	61	65	73	71	70	67	71	67	69	64	57	66
Savannah, GA	46	54	57	62	71	68	65	64	62	58	63	61	55	62
Hilo, HI	56	46	46	42	37	38	44	41	42	43	39	33	37	41

Continued…

Table 7.2 Mean Percentage (%) of Possible Sun, through 2009—continued

City/State	Years	Jan	Feb	Mar	Apr	May	Jun	Jul	Aug	Sep	Oct	Nov	Dec	Annual
Honolulu, HI	46	65	68	72	70	72	74	76	77	77	70	65	63	71
Kahului, HI	37	64	64	64	63	68	72	71	71	73	68	62	63	67
Lihue, HI	57	55	57	59	55	61	63	62	65	68	60	49	49	59
Des Moines, IA	57	51	54	57	56	61	68	72	70	66	62	49	45	59
Sioux City, IA	55	57	56	57	59	61	67	73	70	66	63	51	50	61
Boise, ID	60	40	50	62	68	70	75	87	85	82	69	43	38	64
Pocatello, ID	53	40	53	61	66	67	75	83	81	80	72	47	40	64
Cairo, IL	45	45	50	56	62	65	72	74	75	69	67	51	44	61
Chicago, IL	16	44	49	51	50	58	67	66	62	59	55	38	43	44
Moline, IL	53	48	50	50	53	57	63	68	66	62	58	42	40	55
Peoria, IL	52	47	50	51	55	60	67	69	67	64	61	43	42	56
Springfield, IL	48	48	52	51	56	63	68	71	70	68	63	48	44	59
Evansville, IN	56	42	48	55	60	64	71	73	73	69	65	48	42	59
Fort Wayne, IN	52	46	51	55	60	68	74	75	74	68	62	42	38	59
Indianapolis, IN	53	40	49	50	54	60	65	66	68	65	61	41	38	55
Concordia, KS	34	64	63	63	65	67	76	78	76	70	68	59	57	67
Dodge City, KS	66	67	65	65	68	67	74	79	78	74	73	66	65	70

Topeka, KS	53	56	55	57	58	61	66	71	70	66	64	54	52	61
Wichita, KS	55	58	61	62	64	64	69	76	75	68	64	58	57	65
Louisville, KY	48	41	48	51	56	60	66	67	66	64	61	46	40	56
Paducah, KY	19	45	49	55	62	64	61	71	71	64	65	51	42	58
Lake Charles, LA	19	62	66	74	71	72	78	83	81	78	75	67	59	72
New Orleans, LA	22	46	50	56	62	62	63	58	61	61	64	54	48	57
Shreveport, LA	53	51	56	58	60	63	70	75	74	70	68	60	54	63
Blue Hill, MA	113	46	50	48	49	52	55	57	58	56	55	47	46	52
Boston, MA	61	53	56	57	56	58	63	65	65	63	60	50	52	58
Portland, ME	55	56	59	56	54	54	59	63	63	62	58	48	53	57
Baltimore, MD	40	51	55	56	56	56	62	64	62	60	58	51	49	57
Alpena, MI	37	36	45	53	52	59	63	65	59	51	42	28	28	48
Detroit, MI	31	40	46	52	54	61	66	68	67	61	51	35	31	53
Grand Rapids, MI	36	28	39	46	51	56	62	64	61	54	44	27	23	46
Lansing, MI	42	36	44	49	52	61	65	69	64	59	50	31	29	51
Marquette, MI	24	37	44	50	52	62	66	67	63	57	46	36	35	51
Sault Ste. Marie, MI	55	36	47	55	54	57	58	62	58	45	38	24	27	47

Continued...

Table 7.2 Mean Percentage (%) of Possible Sun, through 2009—continued

City/State	Years	Jan	Feb	Mar	Apr	May	Jun	Jul	Aug	Sep	Oct	Nov	Dec	Annual
Duluth, MN	48	48	53	55	56	57	58	65	61	52	46	35	39	52
Minneapolis–St. Paul, MN	58	53	59	57	58	61	66	72	69	62	55	39	42	58
Columbia, MO	27	50	49	50	55	57	64	67	64	60	59	47	45	56
Kansas City, MO	23	58	55	58	62	61	66	72	67	66	60	49	49	60
St. Louis, MO	37	50	52	54	56	59	66	68	65	63	60	46	43	57
Springfield, MO	57	50	52	56	59	60	65	71	71	67	64	52	48	60
Jackson, MS	33	49	54	60	66	63	70	66	67	65	70	57	49	61
Tupelo, MS	13	53	53	61	73	72	74	75	73	72	62	51	46	64
Billings, MT	57	47	53	62	60	61	64	77	75	68	62	46	44	62
Great Falls, MT	46	49	56	66	62	62	65	79	76	67	61	46	44	61
Helena, MT	55	46	55	61	59	60	64	78	74	67	60	44	42	59
Missoula, MT	57	33	44	54	57	59	63	81	77	69	55	34	29	55
Asheville, NC	32	55	59	61	66	61	62	60	54	56	61	58	55	59
Cape Hatteras, NC	33	48	52	60	67	65	65	65	65	65	60	56	48	60
Charlotte, NC	48	54	58	61	68	67	67	67	65	64	65	58	55	62
Greensboro, NC	68	51	56	60	63	63	64	62	61	62	64	57	53	60

Raleigh, NC	42	52	56	60	63	59	60	60	58	58	60	57	53	58
Wilmington, NC	51	56	59	64	70	67	66	64	62	61	64	63	59	63
Bismarck, ND	63	53	53	58	58	61	64	73	72	65	58	43	47	59
Fargo, ND	54	50	56	58	60	61	62	71	69	60	54	40	43	57
Williston, ND	46	52	57	61	60	62	66	74	74	67	59	46	50	61
Lincoln, NE	40	58	57	57	58	61	69	73	70	66	63	53	52	61
North Platte, NE	54	63	63	62	64	65	71	77	75	72	70	60	61	67
Omaha (North), NE	57	55	53	54	58	61	67	74	70	68	65	51	48	60
Valentine, NE	29	63	62	59	59	62	69	76	76	71	68	60	60	65
Concord, NH	58	52	55	53	53	55	58	62	60	56	53	42	47	54
Mt. Washington, NH	71	32	35	34	34	36	32	30	31	35	38	29	30	33
Atlantic City, NJ	36	50	53	55	56	56	60	61	65	61	59	51	47	56
Albuquerque, NM	63	72	72	73	77	79	83	76	75	79	79	76	71	76
Roswell, NM	7	60	68	75	77	80	83	77	73	72	77	73	71	74
Ely, NV	56	68	68	71	70	72	80	80	81	82	75	67	67	73
Las Vegas, NV	47	77	81	83	87	88	93	88	88	91	87	81	78	85
Reno, NV	45	65	68	75	80	81	85	92	92	91	83	70	64	79

Continued...

Table 7.2 Mean Percentage (%) of Possible Sun, through 2009—continued

City/State	Years	Jan	Feb	Mar	Apr	May	Jun	Jul	Aug	Sep	Oct	Nov	Dec	Annual
Winnemucca, NV	42	51	56	60	66	72	77	86	85	82	74	54	52	68
Albany, NY	64	46	52	54	54	56	60	64	61	57	52	37	39	53
Binghamton, NY	51	37	42	46	50	56	62	64	61	55	49	32	29	49
Buffalo, NY	66	31	38	46	51	56	65	67	64	57	50	29	27	48
New York, Central Park, NY	109	51	55	57	58	61	64	65	64	62	61	52	49	58
Rochester, NY	57	35	41	49	53	59	66	69	66	59	49	31	30	51
Syracuse, NY	60	33	39	47	49	55	59	63	59	54	44	27	25	46
Cleveland, OH	66	31	37	45	53	58	65	67	63	60	52	33	26	49
Columbus, OH	45	36	42	44	51	56	60	60	60	61	56	37	31	50
Dayton, OH	53	40	44	48	52	58	66	66	67	65	59	40	36	53
Toledo, OH	41	41	46	50	52	60	64	65	63	61	54	37	33	52
Oklahoma City, OK	42	60	60	65	68	66	75	79	79	72	70	61	58	68
Tulsa, OK	55	53	56	58	60	60	66	74	73	66	64	56	53	62
Portland, OR	46	28	38	48	52	57	56	69	66	62	44	28	23	48
Allentown, PA	12	43	48	53	46	53	62	57	61	58	57	49	45	53
Avoca, PA	41	41	47	50	53	57	61	62	61	55	52	36	34	51

Harrisburg, PA	53	49	54	58	59	60	65	68	67	62	58	47	44	58
Philadelphia, PA	61	49	53	55	56	57	62	61	62	59	60	52	49	56
Pittsburgh, PA	49	32	36	43	46	50	55	57	56	55	51	36	28	45
Providence, RI	42	56	58	58	57	58	61	63	62	62	61	50	52	58
Charleston, SC	39	56	59	66	72	68	66	67	64	61	63	59	56	63
Columbia, SC	46	55	59	64	70	68	67	67	66	65	67	63	59	64
Greenville–Spartanburg, SC	39	54	57	63	66	62	62	60	61	62	66	58	54	60
Huron, SD	55	57	59	59	61	65	70	76	74	69	63	50	49	63
Rapid City, SD	55	57	60	63	62	60	65	73	74	70	66	55	55	63
Chattanooga, TN	65	43	39	53	61	65	65	62	63	64	63	53	44	57
Knoxville, TN	57	40	47	53	63	64	65	64	63	61	61	49	40	56
Memphis, TN	35	50	54	56	64	69	74	74	75	69	70	58	50	64
Nashville, TN	55	41	47	52	59	60	65	63	63	62	62	50	42	56
Abilene, TX	49	62	64	70	72	70	78	80	78	71	72	67	62	71
Amarillo, TX	61	69	68	72	74	71	78	79	77	73	75	72	67	73
Austin/City, TX	61	49	51	55	54	56	69	74	74	66	64	54	49	60

Continued…

Table 7.2 Mean Percentage (%) of Possible Sun, through 2009—continued

City/State	Years	Jan	Feb	Mar	Apr	May	Jun	Jul	Aug	Sep	Oct	Nov	Dec	Annual
Austin/Bergstrom, TX	64	49	51	55	54	56	69	75	74	66	63	56	49	60
Brownsville, TX	59	41	48	53	58	63	73	80	76	68	65	51	42	60
Corpus Christi, TX	62	44	49	54	56	59	72	79	76	68	67	54	43	60
Dallas-Fort Worth, TX	17	52	54	58	61	57	67	75	73	67	63	57	52	61
El Paso, TX	54	78	82	86	89	90	90	82	81	83	84	83	77	84
Galveston, TX	103	48	51	56	61	67	75	73	71	68	71	59	48	62
Houston, TX	27	45	50	54	58	62	68	70	68	66	64	52	51	59
Lubbock, TX	25	65	66	73	74	71	76	77	76	71	75	69	65	72
Midland–Odessa, TX	22	66	69	73	78	78	81	81	77	77	72	74	65	74
Port Arthur, TX	26	42	52	52	52	64	69	65	63	62	67	57	47	58
San Antonio, TX	53	47	50	57	56	56	67	74	74	67	64	54	48	60
Milford, UT	16	58	64	63	69	73	82	77	79	80	76	62	60	70
Salt Lake City, UT	64	45	54	64	68	72	80	83	82	82	72	53	42	66
Burlington, VT	65	41	48	51	49	56	59	64	60	54	47	31	33	49
Lynchburg, VA	52	52	56	58	62	62	65	62	62	61	62	56	53	59
Norfolk, VA	32	53	56	60	63	62	67	62	62	61	59	56	54	60
Richmond, VA	46	54	58	62	66	66	70	68	66	65	63	59	54	63

Quillayute, WA	30	22	30	34	35	37	35	43	44	47	34	21	19	33
Seattle, WA	31	28	34	42	47	52	49	63	56	53	37	28	23	43
Spokane, WA	48	28	41	55	61	65	67	80	78	72	55	29	23	55
Elkins, WV	11	29	32	39	46	44	48	44	44	45	46	37	28	40
Green Bay, WI	60	49	52	54	56	61	65	66	63	57	48	38	40	54
Madison, WI	50	47	51	52	52	58	64	67	64	60	54	39	40	54
Milwaukee, WI	55	44	47	50	53	60	65	69	66	59	54	39	38	54
Cheyenne, WY	64	64	67	67	63	61	67	69	68	70	69	61	60	66
Lander, WY	50	65	68	70	66	64	72	75	75	72	67	58	61	68
Sheridan, WY	55	57	60	63	60	60	65	75	75	68	62	53	55	63
Guam, PC	41	48	53	57	57	56	51	40	36	38	38	40	38	46
Johnston Island, PC	22	70	74	74	70	74	78	79	76	73	65	59	62	71
Koror, PC	46	53	53	62	61	52	44	45	44	51	46	50	49	51
Majuro, Marshall Island, PC	45	61	64	66	59	58	55	56	61	59	55	53	53	58
Pago Pago, Amer Samoa, PC	41	45	47	47	42	34	35	40	44	51	45	46	47	44

Continued...

Table 7.2 Mean Percentage (%) of Possible Sun, through 2009—continued

City/State	Years	Jan	Feb	Mar	Apr	May	Jun	Jul	Aug	Sep	Oct	Nov	Dec	Annual
Pohnpei, Caroline Island, PC	46	38	41	45	42	41	40	44	44	44	40	39	36	41
Chuuk, E. Caroline Island, PC	46	50	55	55	49	46	43	47	47	44	42	43	42	47
Wake Island, PC	28	68	71	76	75	75	76	72	68	68	68	65	64	71
Yap, W Caroline Island, PC	47	57	59	66	66	63	52	47	45	49	47	52	48	54
San Juan, PR	51	68	70	76	71	64	65	69	68	62	64	59	61	66

Note: Monthly climatic data for the world is available from National Climatic Data Center/National Oceanic and Atmospheric Administration (NCDC/NOAA), February 2012 Report, vol. 65, ISSN 0027–0296.
Source: *Developed from NOAA and NCDC, U.S. Department of Commerce.*

(line H), and the actual effective total collector area installed can be transferred to Table 7.1(a) as annotated. Calculating for estimated system output, we can illustrate the use of Table 7.1(a) line by line for the month of January.

Line A = 31 days/month

Line B = (31 days/month) × (55,395 BTUs/day) = 1.72 MBTU/month

Line C = (31 days/month) × (1478 BTUs/ft²-day) = 45,818 BTUs/ft²-month

Line D = (Effective collector area) × (line C)
= (66.6ft²) × (45,818 BTUs/ft²-month) = 3.05 MBTUs/month

Line E = Percentage of possible sun in Billings, Montana (Table 7.2) = 0.47 (January)

Line F = (Line E) × (line D)
= (0.47) × (3.05 MBTUs/month) = 1.43 MBTUs/month

Line G = Collector system efficiency; Table 4.4(a) = 0.6

Line H = (Line G) × (line F) = (0.6) × (1.43 MBTUs/month) = 0.86 MBTUs/month

Line I = (Line H) ÷ (line B) = (0.86 MBTUs/month) ÷ (1.72 MBTUs/month) = 0.50

This procedure can be followed to determine an estimated system output for each month. The three collector system in Billings, Montana, would yield an estimated 14.19 million BTUs (MBTUs) per year. Because the demand has been determined to be 19.26 MBTUs, the solar DHW system would provide approximately 74.5% of the total hot water needed.

7.1 COST FACTORS

Projecting the cost of energy from conventional fuels with any degree of certainty is a formidable task with fuel source uncertainties and lack of a sensible federal energy plan. The cost of conventional DHW equipment is minimal compared with that of solar DHW equipment, and the cost of installing a typical electric hot water system may be only 10% of the cost of an installed solar DHW system. It is the increasing costs of nonrenewable fuels that make the switch to solar a compelling investment. The average energy inflation rate for consumers in the United States reported by the consumer price index from the U.S. Department of Labor over a 10-year period from 2002 was 7.2% (Chapter 6, Section 6.6). The highest energy inflation rate occurred in 1980 reaching 47.13% with a historic 20-year average of 5.2%.[1] Today's utility bills are squeezing the typical family budget, and tomorrow's utility rates continue to escalate. Until such time that the federal government figures out how to regulate and tax sunshine, the cost of the energy you receive from the sun simply will not change.

Typical costs to heat water electrically are represented in Table 7.3. Local electric rates per kilowatt-hour can be determined either from monthly electric bills (by

---

[1] U.S. Department of Labor Information.

Table 7.3 Typical Domestic Hot Water Electrical Expenses

Water Heated per day (gallon)	Yearly Requirement		Yearly Cost to Heat Water from 40 to 135°F							
	MBTU	kWh	11¢/kWh	12¢/kWh	13¢/kWh	14¢/kWh	15¢/kWh	16¢/kWh	17¢/kWh	18¢/kWh
60	17.3	5067.4	$557.41	$608.09	$658.76	$709.44	$760.11	$810.78	$861.46	$912.13
70	20.2	5916.8	$650.85	$710.02	$769.18	$828.35	$887.52	$946.69	$1005.86	$1065.02
80	23.1	6766.3	$744.29	$811.96	$879.62	$947.28	$1014.95	$1082.61	$1150.27	$1217.93
90	26.0	7615.7	$837.73	$913.88	$990.04	$1066.20	$1142.36	$1218.51	$1294.67	$1370.83
100	28.9	8465.1	$931.16	$1015.81	$1100.46	$1185.11	$1269.77	$1354.42	$1439.07	$1523.72
110	31.8	9314.6	$1024.61	$1117.75	$1210.90	$1304.04	$1397.19	$1490.34	$1583.48	$1676.63
120	34.7	10,164	$1118.04	$1219.68	$1321.32	$1422.96	$1524.60	$1626.24	$1727.88	$1829.52

dividing the cost of quick-recovery water heating by the total number of kilowatt-hours used) or by direct inquiry from the utility company. Remember to include any fuel surcharge or delivery charge in the actual cost as applicable. The information in Table 7.3 is derived from the heat equation in Chapter 3, Section 3.2. For instance,

$$Q = (Wc)\,(Ts - Ti)\,(Cp)\,(8.33)\,(\text{Amount of heat required})$$

$$Q = (60\ \text{gallon/day}) \times (135 - 40°F) \times (1\ \text{BTU/lb–°F}) \times (8.33\ \text{lb/gallon})$$

$$Q = 47,481\ \text{BTUs/day}$$

The amount of heat needed for the year, therefore, is ((47,481 BTU/day)×(365 days/year)) 17.3 MBTUs, which would be the yearly requirement for 60 gallon/day as illustrated in Table 7.3. The energy required for various amounts of hot water and associated costs per kilowatt-hour are represented in Table 7.3 using the same type of calculations.

The cost to heat water electrically for our typical family of four in Billings, Montana, is determined by the same method from which Table 7.3 was derived. From Table 7.1(a), it was determined that this family has a 19.26 MBTU yearly requirement for hot water. Assuming an electric rate of $0.16/kWh, we have a yearly cost of $902.64, whereas

$$(19.26\ \text{MBTU}) \times (1,000,000\ \text{BTU/1 MBTU}) \times (1\ \text{kWh/3414 BTU})$$
$$\times\,(\$0.16/\text{kWh}) = \$902.64$$

This same procedure can be used to determine the yearly cost of all types of fuels using the energy conversion factors discussed in Chapter 3, Section 3.1. The volatility of costs and variable efficiencies, however, must be included to ensure a comprehensive cost analysis.

7.2 EQUIPMENT AND LABOR COSTS

Equipment and labor are, of course, the two major cost factors of a solar DHW system. The cost of the equipment can average around $6000 (at 2012 prices) and the cost of labor for installation can average $4000. Over the past 20 years the average cost of a solar DHW system has more than doubled. This, however, is to be expected because the cost of manufacturing, materials in the building industry, and wages have increased over the years. You also must consider the fact that because of continued printing of our currency, the value of the dollar at present-day values is also worth less. The commercially available flat-plate and evacuated tube collectors are well developed, and dramatic technological advances in improvement will be few.

7.3 MAINTENANCE COSTS

Maintenance costs depend on the durability of system components and the choice of the heat transfer medium. Maintenance on system components should be minimal over a 20-year-life system. In general, heat transfer fluids can cause the highest maintenance cost, especially if they must be changed every 2 or 3 years. Glycol-water solutions are initially inexpensive, normally being one-third the cost of most synthetic hydrocarbons and one-fifth the cost of silicones. When maintenance costs are added to the initial cost of the fluid plus the replacement cost of new fluid every 2 or 3 years, one will find the choice of synthetic hydrocarbons is actually better both monetarily and in terms of operational effectiveness. By using premium grade components and materials, and by the occasional monitoring of temperature and pressure (as applicable), problems (if any) and system efficiency degradation can be kept to a minimum. It is important to remember that maintenance costs will reduce the amount of return from the solar DHW investment.

7.4 OPERATIONAL COSTS

Depending on system type, the only electrically operated components in the solar DHW system are the differential controller and the circulator or pump. The differential controller uses a minimal amount of energy in operation and does not dramatically alter a cost analysis if omitted. Operational costs of a circulator or pump, however, should be considered. The higher the horsepower of an electric circulator or pump, the higher the wattage needed, and therefore the higher the electricity costs. Examples of pump operational costs for various horsepower requirements are illustrated in Table 7.4. A closed-loop system with its low horsepower circulator will cost less to operate than an open-loop system with its higher horsepower pump. For example, if the electric rate is $0.16/kWh and a 1/20 hp circulator is operated an average of 8 h per day year-round, the electric operational cost would be approximately $3.82 ($45.79/12 = $3.82) per month.

7.5 COMPARATIVE ANALYSIS—ELECTRIC UTILITY VERSUS SOLAR DHW

Before we discuss the payback of a solar DHW system, we first will review the cost of heating water with electricity versus the cost of heating water with a solar DHW system. A comparative analysis of a typical solar DHW system versus an electric water heater should substantiate the fact that a solar DHW system is an economically sound and viable investment. This same economic comparison can be conducted using any of the other energy sources at their current costs.

As an example, the solar contribution of a solar DHW system for the typical family of four in Billings, Montana, was determined in Table 7.1(a) to be **14.19 MBTUs** per year, which would provide approximately 74% of their hot water requirements. At $0.16/kWh, it would cost **$665.22** per year to heat the same amount of water with conventional electric heating. This cost is calculated from the following:

$$(14,190,000 \text{ BTUs}) \times (1 \text{ kWh}/3413 \text{ BTUs}) \times (\$0.16/\text{kWh}) = \$665.22$$

Using a 1/20 hp circulator, the solar operational cost can be determined from Table 7.4 to be **$45.79** per year. As an additional conservative measure, we shall

Table 7.4 Typical Pump Operational Costs

Horsepower of Pump or Circulator	Kilo-watts Required	Annual kWh (based upon 8 h/day)	Annual Costs at Various Electric Rates							
			11¢/kWh	12¢/kWh	13¢/kWh	14¢/kWh	15¢/kWh	16¢/kWh	17¢/kWh	18¢/kWh
1/25	0.085	248.2	$27.30	$29.78	$32.27	$34.75	$37.23	$39.71	$42.19	$44.68
1/20	0.098	286.2	$31.48	$34.34	$37.21	$40.07	$42.93	$45.79	$48.65	$51.52
1/12	0.185	540.2	$59.42	$64.82	$70.23	$75.63	$81.03	$86.43	$91.83	$97.24
1/4	0.420	1226.4	$134.90	$147.17	$159.43	$171.70	$183.96	$196.22	$208.49	$220.75
1/3	0.530	1547.6	$170.24	$185.71	$201.19	$216.66	$232.14	$247.62	$263.09	$278.57

assume an average **$150** yearly maintenance fee for the solar DHW system. Yearly maintenance fees often are offered by the dealers or installers to check for proper system operation and pH integrity of the transfer fluid. We therefore have the first yearly conventional cost to heat water in Table 7.5, column 1 at $0.16/kWh and a solar operational and maintenance cost of $195.79 in Table 7.5, column 3. The cost and savings comparison realized from using the sun's energy versus a conventional form of energy is illustrated for each year in Table 7.5. At this point, the actual cost of the solar energy system has not been included, but this cost will be addressed during the discussion of the system's actual payback time period.

Table 7.5 Example of Energy Costs and Savings Realized (System Output of 14.19 MBTUs at $0.16/kWh)

Years	Conventional Electrical DHW Fuel Costs at 5% per year Inflation		Solar Operational and Maintenance Costs at 5% per year Inflation		Savings Realized from Solar versus Conventional Electric	
	(1) Yearly	(2) Cumulative	(3) Yearly	(4) Cumulative	(1)−(3) Yearly	(2)−(4) Cumulative
1	$665.22	$665.22	$195.79	$195.79	$469.43	$469.43
2	698.48	1363.70	205.58	401.37	492.90	962.33
3	733.41	2097.11	215.86	617.23	517.55	1479.88
4	770.08	2867.18	226.65	843.88	543.42	2023.30
5	808.58	3675.76	237.98	1081.86	570.60	2593.90
6	849.01	4524.77	249.88	1331.75	599.12	3193.02
7	891.46	5416.23	262.38	1594.12	629.08	3822.10
8	936.03	6352.26	275.50	1869.62	660.54	4482.64
9	982.83	7335.09	289.27	2158.89	693.56	5176.20
10	1031.97	8367.07	303.73	2462.63	728.24	5904.44
11	1083.57	9450.64	318.92	2781.55	764.65	6669.09
12	1137.75	10,588.39	334.87	3116.41	802.88	7471.98
13	1194.64	11,783.03	351.61	3468.02	843.03	8315.01
14	1254.37	13,037.40	369.19	3837.22	885.18	9200.19
15	1317.09	14,354.49	387.65	4224.87	929.44	10,129.63
16	1382.94	15,737.44	407.03	4631.90	975.91	11,105.54
17	1452.09	17,189.53	427.39	5059.29	1024.71	12,130.24
18	1524.70	18,714.22	448.75	5508.04	1075.94	13,206.19
19	1600.93	20,315.16	471.19	5979.23	1129.74	14,335.92
20	1680.98	21,996.13	494.75	6473.98	1186.23	15,522.15
	Total	$21,996.13	Total	$6473.98	Total	$15,522.15

The cost comparison shown in Table 7.5 is actually a bit conservative regarding actual electric water heating costs. The table uses a conservative energy inflation rate of only 5%, whereas fossil fuels normally continue to have higher rates of inflation. In addition, the life span of most conventional hot water heaters does not exceed 20 years at which time they have to be replaced.

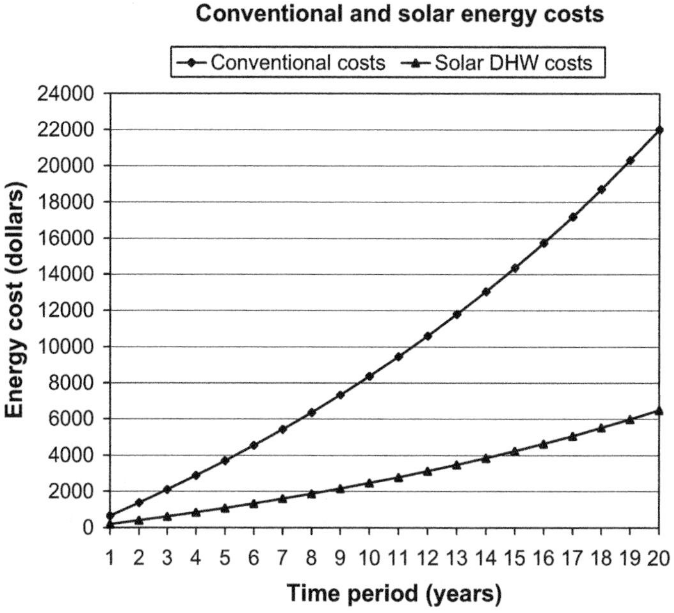

Conventional and solar energy costs

FIGURE 7.1 Cost of producing hot water—conventional electricity versus solar.

Salvage value of the solar DHW components and possible component replacement with the exception of a yearly maintenance fee also is not considered. By omitting these factors, however, we should have a relatively complementary view of energy costs. The costs of producing hot water by conventional electric means versus solar is presented in Table 7.5 and is graphically illustrated in Figure 7.1. This graph shows a $15,522.15 difference in the cost of heating water with electricity versus a solar hot water system over a 20-year period. Next we need to consider the actual cost of the solar equipment, which will increase the payback period, as well as the tax incentives available, which will help reduce the actual costs, decreasing the payback period.

7.6 PAYBACK ANALYSIS BEFORE TAX CREDIT INCENTIVES

Let's examine a typical solar DHW system proposal, including cost of materials and installation charges. Some solar dealers will provide a price separation for components and labor. Other dealers include both factors as a total package installed.

7.6.1 Example: Proposed System

EQUIPMENT

This actual quotation proposes a closed-loop antifreeze solar hot water system, utilizing the following components:

1. Two Wagner and Co. Euro C20 AR-M flat-plate collectors with silver frames
2. 80-gallon Caleffi solar storage tank with electric element
3. One Stiebel Eltron Flowstar solar pump station
4. One Stiebel Eltron Som 6 differential temperature controller
5. Propolyene-glycol 40% noncorrosive solution freeze-resistant transfer fluid
6. Caleffi series 521 antiscald mixing valve

INSTALLATION

Collectors will be flush mounted on the south-facing roof designed for primary solar domestic water heating with backup coming from the electric element in the storage tank. On a clear day, the system will produce a 60–70 °F temperature rise in the solar tank and will provide a significant portion of the DHW requirements. The system will produce more than 14.19 million BTUs of clean renewable heat energy annually. Two half-inch copper pipes insulated with foam will be installed between the rooftop collectors and the storage tank. Pipes will exit the collectors, penetrate into the attic, penetrate down through the first and second floor closets, and then penetrate into the basement and traverse the basement ceiling over to the solar storage tank.

(Cost of system installed) = $10,891.

Before we consider the savings introduced with the energy tax credits available, we will graphically illustrate the payback of the solar DHW system versus a conventional electric hot water system by including the equipment and operational costs of both conventional electric and solar DHW systems. We will assume an estimated initial cost of the conventional energy equipment to be $800. This means that the conventional cost curve originally shown in Figure 7.1 will cross the *y*-axis or ordinate at $800 shifting that curve slightly higher. In addition, the solar DHW cost curve then will cross the *y*-axis or ordinate at $10,891 shifting that curve higher as well. The point at which the solar cost curve intersects the conventional cost curve illustrates the number of years until the costs equate with one another. These added changes are illustrated in Figure 7.2.

Without reducing the cost of the system with the available federal and state tax credits, Figure 7.2 shows that the payback period for solar DHW versus heating water with electricity is approximately 15 years, 9 months. It is the federal and state tax credits that make an important difference in the payback period of a solar alternative energy system.

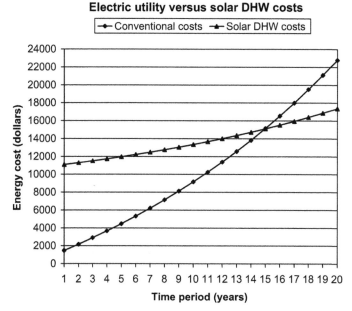

FIGURE 7.2 Cost of equipment and operation—conventional electric versus solar.

7.7 SOLAR ENERGY TAX CREDITS

As mentioned previously in Chapter 6, Section 6.4, the federal government and many states have enacted various income tax credits, rebates, and property tax credit incentives to make solar energy a more attractive energy alternative by helping to lower the cost of equipment and installation. Review your particular state's tax incentive requirements to obtain a rebate or refund. Do not forget that some state rebates for solar DHW installations may require the installer to be a licensed master plumber, a master oil burner technician, or a propane and natural gas technician. Some states may also require an energy audit performed as a tax incentive rebate prerequisite. As mentioned previously, information regarding state tax incentives is available from a Database of State Incentives for Renewable Energy, which is available online (www.dsireusa.org/solar/) for each state. The state rebate savings alone could determine whether it is worth your time and needed skills to install your own system versus a certified installer. Whether you install a system yourself or have it installed by a certified and licensed dealer, you will still save money over a period of time by using solar as an alternative energy. In particular, we shall detail cost savings and review the payback periods for these investments with the addition of tax incentives.

7.8 PAYBACK ANALYSIS WITH TAX CREDIT INCENTIVES

Now let's include the federal tax credit into the previous graph of Figure 7.2 to reflect the actual cost of the solar DHW system after subtracting the available tax credit from the cost of the system. We also will include an average state tax credit of $1000 for our example to provide a conservative payback estimate:

$$\text{Solar DHW system cost} = +\$10, 891.00$$
$$\text{Federal tax credit } (30\%) = -\$ \quad 3267.30$$
$$\text{Actual system cost after federal tax credit} = \$ \quad 7623.70$$
$$\text{State tax credit } (i.e., \text{ billings, MT}) = -\$ \quad 1000.00$$
$$\text{Final system cost after tax credits} = \$ \quad 6623.70$$

Our new graph, Figure 7.3, illustrates the payback period with the added tax credit incentives.

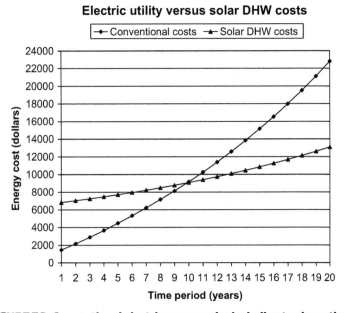

FIGURE 7.3 Conventional electric versus solar including tax incentives.

The payback and break-even point realized from savings after federal and state tax credits can be seen in Figure 7.3 to have been reduced from approximately 15 years, 9 months to 10 years, 11 months. The total actual savings after tax credits are the savings realized from the use of solar energy ($15,522.10) minus the estimated cost of the solar energy system installation ($6623.70) plus the estimated cost of the conventional energy equipment ($800.00) for a cumulative savings of $9698.40 over 20 years. If the future worth of savings is considered throughout a 20-year period for each of the yearly savings shown in Table 7.6, column 3, assuming a conservative 5% energy inflation rate, the resultant future worth savings can be determined using Chapter 6, Section 6.1, Eqn (6.2) and the single payment compound amount factors (SPCAF) in Table 6.1 for each year as follows:

Where:

S = future worth of money,

P = savings for a particular year (Table 7.6, column 3),

(5%—SPCAF) = single payment compound amount factor (Table 6.1 factor)

$S_{20} = P$ (5%—20 SPCAF) = 469.43 (2.6533) = $1245.54
$S_{19} = P$ (5%—19 SPCAF) = 492.90 (2.5269) = $1245.51
$S_{18} = P$ (5%—18 SPCAF) = 517.55 (2.4066) = $1245.54
$S_{17} = P$ (5%—17 SPCAF) = 543.42 (2.2920) = $1245.52
$\vdots$ $\qquad$ $\vdots$ $\qquad$ $\vdots$ $\qquad$ $\vdots$
$S_1 = P$ (5%—1 SPCAF) = 1186.23 (1.0500) = $1245.54

Developing the future worth of savings from the information included in Table 7.5, we have the yearly future worth of savings as shown in Table 7.6. At this point, we should determine whether it would be wiser to invest in the cost of the solar energy system or keep the money in a bank savings account. So what is our money worth if we simply put the cost of the solar DHW system into a regular savings account? Using 2012 interest rates, we are only going to get a 1% interest rate (at best) and more likely only 0.5% interest for our money in a bank. But, let's calculate which option would be the better investment even at a 1% interest rate of return from a savings account.

The savings bank investment in comparison with the solar DHW investment, illustrated in the following example, clearly illustrates that the investment in solar DHW provides a positive cash savings.

Example of Savings Bank versus Solar DHW Investments

Bank (Savings)

Future worth of $6623.70 savings for 20 years @ 1% compounded interest $S_{20} = P$ (1%—20 SPCAF) = ($6623.70) × (1.220)	= +$8080.91
Less expense to heat domestic water (over 20 years)	= −$21,996.13
Less estimated conventional equipment cost (year 1)	= −$800.00
Actual money lost from bank savings of $6623.70 over 20 years	= −$14,715.22

Solar DHW Investment

Future worth of energy savings for 20 years @ inflation rate of 5% compounded interest (Table 7.6, column 6) total over 20 Years	= +$24,910.66
Less expense for solar energy system operation and maintenance over 20 years (Table 7.5, column 4)	= −$6473.98
Less expense for solar equipment and labor installation after federal and state tax credits	= −$6623.70
Actual money saved by installing a solar DHW system	= +$11,812.98

Table 7.6 Example of Energy Costs and Future Worth of Savings Realized

Years	Conventional DHW Fuel Costs at 5% per year Energy Inflation		Savings Realized from Solar versus Conventional Electric		Future Worth of Savings through 20 years at 5% Compounded Inflation	
	(1) Yearly	(2) Cumulative	(3) Yearly	(4) Cumulative	(5) Yearly	(6) Cumulative
1	$665.22	$665.22	$469.43	$469.43	$1245.54	$1245.54
2	698.48	1363.70	492.90	962.33	$1245.51	$2491.05
3	733.41	2097.11	517.55	1479.88	$1245.54	$3736.59
4	770.08	2867.18	543.42	2023.30	$1245.52	$4982.11
5	808.58	3675.76	570.60	2593.90	$1245.56	$6227.67
6	849.01	4524.77	599.12	3193.02	$1245.51	$7473.18
7	891.46	5416.23	629.08	3822.10	$1245.51	$8718.69
8	936.03	6352.26	660.54	4482.64	$1245.51	$9964.20
9	982.83	7335.09	693.56	5176.20	$1245.56	$11,209.76
10	1031.97	8367.07	728.24	5904.44	$1245.51	$12,455.27
11	1083.57	9450.64	764.65	6669.09	$1245.54	$13,700.81
12	1137.75	10,588.39	802.88	7471.98	$1245.51	$14,946.32
13	1194.64	11,783.03	843.03	8315.01	$1245.58	$16,191.90
14	1254.37	13,037.40	885.18	9200.19	$1245.54	$17,437.44
15	1317.09	14,354.49	929.44	10,129.63	$1245.54	$18,682.98
16	1382.94	15,737.44	975.91	11,105.54	$1245.55	$19,928.53
17	1452.09	17,189.53	1024.71	12,130.24	$1245.54	$21,174.07
18	1524.70	18,714.22	1075.94	13,206.19	$1245.51	$22,419.58
19	1600.93	20,315.16	1129.74	14,335.92	$1245.54	$23,665.12
20	1680.98	21,996.13	1186.23	15,522.15	$1245.54	$24,910.66
	Total	$21,996.13	Total	$15,522.15	Total	$24,910.66

The solar investment also provides a better use of *cash flow* throughout the 20-year period. This example illustrates that purchasing a solar DHW system produces a $26,528.20 difference by adding the gain in money saved by heating your water with solar energy versus putting that same investment in the bank and losing $14,715.22 because of energy inflation over the same 20-year period. The solar DHW cost is self-liquidating, in that once the system cost has been repaid, only maintenance costs and electricity costs for pump operation remain.

Every solar DHW system represents an individual case, and an economic evaluation of each individual case will indicate that solar energy is an excellent first application in using the sun's energy to heat domestic water rather than using conventional fuels. The example costs and calculations used throughout this chapter have been conservative, representing this type of investment. It also is assumed that the system is exempt from sales tax and exempt from property tax, which is true in some but not all states. In most situations, the return on investment is less than 10 years. For instance, if we had used an inflation factor of 8% versus 5% in example (Table 7.5), the cost to heat water with electricity over 20 years would be $30442 versus $21996, the cost for solar DHW operation and maintenance would be $8960 versus $6474, and the savings realized using solar would be $21,482 versus $15,522. In addition, comparisons to using fuel sources other than electricity, such as oil, may produce even shorter payback times because of volatile fuel prices and efficiency losses.

Chapter | eight

The Economics of Solar Photovoltaic Systems

This chapter explains and illustrates the economic advantage of using solar energy to supplement your electric utility energy demands. In particular, examples of cost savings and payback periods for photovoltaic (PV) systems will be reviewed as potential investments. In the case of evaluating PV systems versus a solar domestic hot water (DHW) system, it is not a matter of determining comparative costs of several energy sources, such as oil, natural gas, propane, or electricity, to meet the necessary energy requirements. Rather, it is a matter of determining the comparative cost of one energy source, electricity, from your power utility company and using a PV energy system to reduce the cost of that energy purchased. In essence, a PV system generates individual residential power, supplementing all of your electrical energy requirements.

Magazine articles printed more than 20 years ago argued the case that solar PV systems would need to achieve a cost of $1.00/W to economically rival conventional power generation in the 1990s and beyond. By early 2006, the average cost per installed Watt for a residential-size PV electric system was between $7.50 and $9.50, including solar PV panels, inverters to convert direct current (DC) to alternating current (AC), collector mounts, and electrical components. By 2012, these costs had been reduced to approximately $4.00–6.00/W installed. To date, despite the failure to achieve that $1.00/W goal, the continually increasing costs of energy and the return of federal and state tax incentives have made such systems economically viable.

8.1 COST FACTORS

The cost of electricity from state to state and from one region to another depends on the generation energy source. Current average retail prices of electricity for each state can be found online using the U.S. Energy Information Administration website at http://www.eia.gov/electricity/monthly. Your actual local electric rates per kilowatt-hour and annual costs can be determined directly from your current monthly electric utility bill. Simply add your 12 monthly bills to determine your yearly expense for electricity. Table 8.1 illustrates that the cost of electricity can escalate quickly with only slight increases in utility rates. As noted in Chapter 6, Section 6.6, the average energy inflation rate over a 10-year period from 2002 was

Table 8.1 Annual Cost of Electricity at Various Rates

Annual Household Electric Demand (kWh)	$0.11/kWh	$0.12/kWh	$0.13/kWh	$0.14/kWh	$0.15/kWh	$0.16/kWh	$0.17/kWh	$0.18/kWh
6000	$660.00	$720.00	$780.00	$840.00	$900.00	$960.00	$1020.00	$1080.00
7000	$770.00	$840.00	$910.00	$980.00	$1050.00	$1120.00	$1190.00	$1280.00
8000	$880.00	$960.00	$1040.00	$1120.00	$1200.00	$1280.00	$1360.00	$1440.00
9000	$990.00	$1080.00	$1170.00	$1260.00	$1350.00	$1440.00	$1530.00	$1620.00
10,000	$1100.00	$1200.00	$1300.00	$1400.00	$1500.00	$1600.00	$1700.00	$1800.00

approximately 7.2% annually. Throughout our discussion, we have taken a more conservative approach using 5% as the annual energy inflation factor.

The amount of electricity produced by a PV array can be determined by the same methods that were used in determining PV array sizing as summarized in Chapter 5, Section 5.4, Table 5.3. Assume that we are limited by the area available on a roof and that 18 Canadian solar PV modules (Model CS6P-240M) are to be installed on a true south–facing house at a 30-degree roof angle in Portland, Maine, where the number of peak sun hours per Chapter 5, Section 5.4, Table 5.2 is 4.51. Such a system would be considered a 4.32-kilowatt system using the Standard Test Conditions plate rating (Chapter 5, Section 5.3) of each panel, as follows:

$$(18 \text{ PV modules}) \times (240 \text{ W/module}) = 4320 \text{ W} = \textbf{4.32 kW}$$

In accordance with specifications from Chapter 5, Table 5.1, the PVUSA Test Conditions/California Energy Commission (PTC/CEC) output rating for the Canadian solar module is 212 W for each panel. The total output from this PV array per day therefore would be as follows:

$$(18 \text{ modules}) \times (212 \text{ W/module}) \times (4.51 \text{ h/day}) = 17,210 \text{ Wh/day}$$
$$= 17.2 \text{ kWh/day}$$
$$\text{or } (17.2 \text{ kWh/day}) \times (365 \text{ days/year}) = 6278 \text{ kWh/year}$$

Assuming an 85% efficiency loss factor or derating factor as addressed in Chapter 5, Section 5.4 and assuming the collector orientation and tilt parameters per Chapter 2, Sections 2.4.1 and 2.4.2 respectively, have been met, the actual energy output would be approximately as follows:

$$(6278 \text{ kWh/year}) \times (0.85) = \textbf{5336 kWh/year}$$

The annual cost of supplying that amount of energy for various electricity rates is illustrated in Table 8.2, demonstrating once again that slight increases in utility rates can lead to significant increases in yearly expenses.

Table 8.2 Yearly Cost of Electricity at Various Rates

Yearly PV Array Output (kWh)	$0.11/ kWh	$0.12/ kWh	$0.13/ kWh	$0.14/ kWh	$0.15/ kWh	$0.16/ kWh	$0.17/ kWh	$0.18/ kWh
5336	$586.96	$640.32	$693.68	$750.40	$800.40	$853.76	$907.12	$960.48

Instead of manually calculating the power output from a specific number of PV modules as illustrated, you can determine the output of a specified PV module array with an online Grid Data calculator provided by the National Renewable Energy Laboratory (NREL). The **PVWatts™ Grid Data Calculator** is available at the Renewable Resource Data Center on the NREL website (www.nrel.gov/rredc).

For example, if you go to the NREL website and enter a DC rating of 4.32 kWh, with a derating or efficiency loss factor of 0.85, an array type as fixed tilt at 30°, and a true south–facing site at 180°, you will receive an annual energy result of **5643 kWh/year**. This result corresponds fairly closely with the manual calculation of **5336 kWh** derived earlier. Either one of these methods will provide you an approximate output based on your particular situation. Because links to various online website calculators can change over time, however, manual calculations always can be performed without computer and Internet assistance.

8.2 EQUIPMENT AND LABOR COSTS

Just like solar DHW systems, equipment and installation costs are the two major cost factors for PV systems. By 2012, prices of solar PV modules actually dropped significantly to roughly $2/W for the PV module itself and $4/W for installation, inverter, and wiring costs. The only potential component replacement during an anticipated 25-year life of a solar PV system is the replacement of an inverter. Based on estimates from the Department of Energy Solar Energy Technologies Program, replacement costs of inverters average approximately $300/kW over a 10-year period. These costs are subject to change as the cost of inverters continues to decline as demand increases and technology advances. Some manufacturers already provide a 15-plus-year warranty for inverters with a fail period after the first 20 plus years.

PV systems provide a way to reduce fossil fuel energy consumption while locking in your electric rate below the electrical grid average for a duration of 25 years plus. During that time period, most manufacturers specify an output panel degradation drop at 0.5–1.0% per year with most panels guaranteed not to drop below 80% in 25 years. Keep in mind that if a systematic degradation initiates warranty claims against a large number of collector manufacturers, solvency can become a concern. A company that has made promises it cannot keep could go bankrupt within that time period, nullifying their warranty. Do not despair, however, because some manufacturers insure their warranties with a separate insurance company in case of insolvency. So check with the dealer or installer and ask about the manufacturer and their guarantee of a product warranty.

Unlike solar DHW systems, there are relatively no additional costs for operation or maintenance of a PV system. Depending on the number of PV modules, the overall costs of PV systems normally are greater than solar hot water systems because the manufacturing costs of the modules and electrical inverters are more expensive. This results in a longer return on investment. Although PV module prices have decreased over the past few years, the cost of labor for installation of the modules, wiring, and DC inverters have not. Cost savings are greater, however, over the same period because the demand and use of household electricity is greater than the energy demand for hot water, there are less system losses, there are no additional monthly costs for operation or maintenance, and there is a greater percentage of use. As a result, PV solar energy systems may have only a slightly longer payback period than solar DHW systems.

8.3 COMPARATIVE ANALYSIS—ELECTRIC UTILITY VERSUS SOLAR PV

Table 8.3 illustrates a comparison of utility-provided electricity costs with the cost savings generated by a PV array producing an estimated 5336 kilowatts per year. This table assumes a 5% electricity inflation factor and a 1% module output degradation factor for each year over a 20-year period to provide a conservative savings estimate. The table does not include the additional cost of an inverter should one happen to fail during the 20-year period.

Table 8.3 Example of Utility-Provided Electricity Costs at $0.16/kWh (from Table 8.2) Compared with Cost Savings of 5336 kW Generated by a PV Array

	Cost of Utility-Provided Electricity at 5% per year Energy Inflation		Solar PV Module Losses at 1% Output Degradation per year		Savings Realized from Solar PV versus Electric Utility	
Years	(1) Yearly	(2) Cumulative	(3) Yearly	(4) Cumulative	(5) Yearly (1)-(3)	(6) Cumulative (2)-(4)
1	$853.76	$853.76	$8.54	$8.54	$845.22	$845.22
2	$896.45	$1750.21	$17.93	$26.47	$878.52	$1723.74
3	$941.27	$2691.48	$28.24	$54.71	$913.03	$2636.77
4	$988.33	$3679.81	$39.53	$94.24	$948.80	$3585.57
5	$1037.75	$4717.56	$51.89	$146.13	$985.86	$4571.43
6	$1089.64	$5807.20	$65.38	$211.51	$1024.26	$5595.69
7	$1144.12	$6951.32	$80.09	$291.60	$1064.03	$6659.72
8	$1201.33	$8152.65	$96.11	$387.71	$1105.22	$7764.94
9	$1261.39	$9414.04	$113.53	$501.24	$1147.86	$8912.80
10	$1324.46	$10,738.50	$132.45	$633.69	$1192.01	$10,104.81
11	$1390.69	$12,129.19	$152.98	$786.67	$1237.71	$11,342.52
12	$1460.22	$13,589.41	$175.23	$961.90	$1284.99	$12,627.51
13	$1533.23	$15,122.64	$199.32	$1161.22	$1333.91	$13,961.42
14	$1609.89	$16,732.53	$225.38	$1386.60	$1384.51	$15,345.93
15	$1690.39	$18,422.91	$253.56	$1640.16	$1436.83	$16,782.75
16	$1774.91	$20,197.82	$283.99	$1924.15	$1490.92	$18,273.67
17	$1863.65	$22,061.47	$316.82	$2240.97	$1546.83	$19,820.50
18	$1956.83	$24,018.30	$352.23	$2593.20	$1604.60	$21,425.10
19	$2054.68	$26,072.98	$390.39	$2983.59	$1664.29	$23,089.39
20	$2157.41	$28,230.39	$431.48	$3415.07	$1725.93	$24,815.32
	Total	$28,230.39	Total	$3415.07	Total	$24,815.32

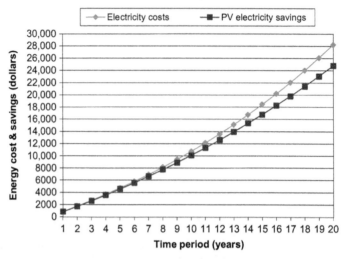

FIGURE 8.1 Power utility electricity costs versus PV savings.

Table 8.3 is graphically illustrated in Figure 8.1 and shows the cost of electricity over a 20-year time period to be $28,230.39 at $0.16/kWh with an energy inflation rate of 5%. The actual savings is reduced by $3415.07 due to a PV module output reduction for each year over 20 years, resulting in an actual cumulative energy savings of $24,815.32.

As discussed in Chapter 5, a PV array can be established using either a single "string inverter" for a series of PV modules or individual "microinverters" for each PV module depending on shading issues at the site. Shading conditions or a defect or failure of one or more PV modules could shut down an entire "string" of modules if there is insufficient power to supply a single "string inverter." We therefore shall examine the cost and payback of three different inverter configurations based on two rows of nine PV modules as shown in the rendering of Figure 8.2.

FIGURE 8.2 Proposed 18-panel PV array, New England region. Rendering courtesy of ReVision Energy. (For color version of this figure, the reader is referred to the online version of this book.)

Because of potential shading problems and possible residual snow conditions in certain regions that could cover the bottom row of PV modules, three types of PV system scenarios should be considered. Some solar dealers or installers will provide a price separation for components and labor. Other dealers include both factors as a total package installed. We will compare actual cost quotations (prepared April, 2012) for three examples of DC to AC inverter combinations.

Example 1: 18 PV modules using microinverters for all 18 modules
Example 2: One single "string inverter" for all 18 PV modules
Example 3: Two "string inverters" with one inverter for each string of nine PV modules

8.3.1 Example 1: Configuration with Microinverters

EQUIPMENT
This sample quotation proposes a 4.32-kilowatt grid-tied array coupled with Enphase Energy Microinverters utilizing the following components:

- Eighteen 240 W Canadian solar PV panels
- Eighteen Enphase M-210 microinverters
- Lifetime subscription to Enphase Enlighten monitoring system
- IronRidge aluminum flush roof mounting system
- All hardware, disconnects, cable, and labor to provide code-compliant North American Board of Certified Energy Practitioners (NABCEP)–certified installation

PERFORMANCE
Collectors will be flush mounted on the south-facing roof on an IronRidge aluminum mounting system and the total array area will be approximately 312 square feet. The system will produce approximately 5336 Kwh of clean, renewable electricity annually and roughly offset 7182 lb of carbon-dioxide emissions annually. Whenever sun shines on the solar PV modules, DC electricity will be generated. The DC electricity from each PV module is converted to AC electricity by the individual Enphase inverters, affixed to the underside of each module. The advantage of microinverters is that the output of the rest of the array is not affected if a portion of panels is shaded. The AC electricity that is created by the inverters then will feed directly into the building's load center. Any loads operating while the sun is shining will be fed directly by the solar electricity. The local utility company will record the amount of electricity that is fed into the grid. If there is more electricity generated by the sun than being used in the house, the second meter will record the amount, creating credit on the next utility bill. The surplus in electricity can be "banked" from month to month for up to one year.

(Cost of system installed) = $19,012.00

8.3.2 Example 2: Configuration with One String Inverter

EQUIPMENT

This sample quotation proposes a 4.32-kilowatt grid-tied array coupled with one single Solectria inverter for two rows of 9 PV modules utilizing the following components:

- Eighteen 240 Watt Canadian solar PV panels
- One Solectria PVI4000 grid-tied inverter
- IronRidge aluminum flush roof mounting system
- All hardware, disconnects, cable, and labor to provide code-compliant NABCEP-certified installation

PERFORMANCE

Collectors will be flush mounted on the south facing roof on an IronRidge aluminum mounting system and the total array area will be approximately 312 square feet. The system will produce approximately 5336 kWh of clean, renewable electricity annually and roughly offset 7182 lb of carbon dioxide emissions annually. Whenever sun shines on the solar PV modules, DC electricity will be generated. The DC electricity will be cabled in conduit to the inverter in the basement. The inverter, which converts direct current to AC, then will feed directly into the electric panel. The local utility company will record the amount of electricity that is fed into the grid. If there is more electricity generated by the sun than being used in the house, the second meter will record the amount, creating credit on the next utility bill. The surplus in electricity can be "banked" from month to month for up to 1 year.

(Cost of system installed) = $16,608.00

8.3.3 Example 3: Configuration with Two String Inverters

EQUIPMENT

This sample quotation proposes a 4.32-kilowatt grid-tied array coupled with two coupled Sunny Boy SMA 2000HF inverters each managing one row of nine PV modules utilizing the following components:

- Eighteen 240 W Canadian solar PV panels
- Two Sunny Boy SMA 2000HF grid-tied inverters
- IronRidge aluminum flush roof mounting system
- All hardware, disconnects, cable, and labor to provide code-compliant NABCEP-certified installation

PERFORMANCE

The performance for the system in Example 3 is the same as in Example 2.

(Cost of system installed) = $18,594.00

Using the preceding examples, let's determine whether the purchase of a PV system is a cost-effective investment for your home.

8.4 PAYBACK ANALYSIS BEFORE TAX CREDIT INCENTIVES

Let's examine the three preceding example quotations for a PV system, including the cost of equipment and installation charges. Before we consider the savings introduced with the energy tax credits available, we will graphically illustrate the payback and savings of the PV system with respect to the cost of purchasing electricity from your power utility company without any tax credits. Because there is a PV module electrical output degradation as mentioned previously in the amount of 0.5–1% per year, we assume these systems will provide a slightly reduced estimated savings of $24,815.32 over a 20-year period as shown in Figure 8.1 in lieu of the total utility costs of $28,230.39. Unlike a solar DHW system, no operating costs and minimal maintenance programs are required, excluding unexpected defects or damage. Figure 8.3 illustrates a one-time cost (represented as a horizontal line) for each example. The point at which the electricity cost curve crosses each of the PV system examples illustrates the "payback period" (the number of years at which the costs equate with one another) as shown in Figure 8.3. System Example 1 shows a break-even payback period of approximately 16 years, 6 months. System Example 2 shows a break-even payback period of approximately 14 years, 11 months. And system Example 3 shows a break-even payback period of approximately 16 years, 3 months. Without reducing the cost of each system with the available federal and state tax credits, Figure 8.3 illustrates an average break-even payback period of approximately 16 years based on the quotation costs provided. The federal and state tax credits make a very important difference in the payback period.

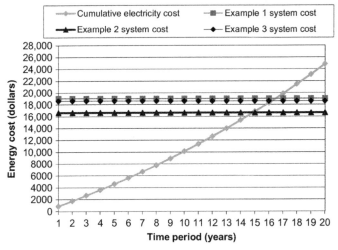

FIGURE 8.3 Conventional electric utility and solar energy costs before tax credits.

8.5 SOLAR ENERGY TAX CREDITS

As mentioned in Chapter 6, a 30% federal residential energy tax credit is available through 2016 with no limit toward the cost of a PV system. This tax credit includes all labor and equipment costs and can be carried forward to future tax years if you cannot take the full credit in the year the system was installed. Be sure to check the availability of these tax credits before purchasing a solar energy system to determine your true costs and payback period. As mentioned in Chapter 6, Section 6.4, some of the state tax rebate incentives require PV installers to have a master's electrician's license and be NABCEP certified or to work with someone who is NABCEP certified.

Some states also require the performance of an energy audit as a tax incentive rebate prerequisite. As mentioned, a database of state tax incentives for renewable energy is available online at www.dsireusa.org/solar/. The state rebate savings alone could justify the additional costs of professional certified installation compared with a do-it-yourself installation. In addition, it is important that you be aware of the electrical safety standards pertaining to the National Electrical Codes for installation of grid-tied systems. Whether you install a system yourself or have it installed by a certified and licensed dealer, you will save money over a period of time by supplementing with solar energy. In particular, we shall now detail cost savings and projected payback periods for these investments with the addition of tax incentives.

8.6 PAYBACK ANALYSIS WITH TAX CREDIT INCENTIVES

With the generous federal tax credits and some state rebates, grid-tied PV systems can be an excellent investment. Now let's include the federal tax credit into the previous graph of Figure 8.3 to reflect the actual cost of the solar PV system after subtracting the available tax credits from the cost of the system. We also will include a possible state tax credit of $2000, which depends on the location of the proposed system.

Example 1: Configuration with Microinverters

(Cost of system installed)	= $ 19,012.00
Federal tax credit (30%)	= – $ 5703.60
Actual system cost after fed.tax credit	= $ 13,308.40
State Tax Credit	= –$ 2000.00
System Cost After Tax Credits	= $ 11,308.40

Example 2: Configuration with One String Inverter

(Cost of system installed)	= $ 16,608.00
Federal Tax Credit (30%)	= – $ 4982.40
Actual System Cost After Fed.Tax Credit	= $ 11,625.60
State Tax Credit	= –$ 2000.00
System Cost After Tax Credits	= $ 9,625.60

Example 3: Configuration with Two String Inverters

(Cost of system installed)	= $ 18,594.00
Federal Tax Credit (30%)	= – $ 5578.20
Actual System Cost After Fed.Tax Credit	= $ 13,015.80
State Tax Credit	= –$ 2000.00
System Cost After Tax Credits	= $ 11,015.80

Our new graph, Figure 8.4, illustrates the payback period with the added federal and state tax credit incentives.

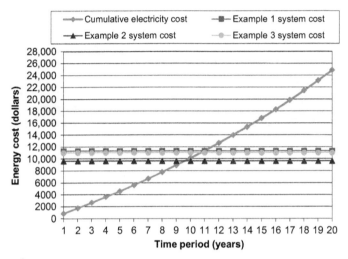

FIGURE 8.4 **Conventional electric utility and solar energy costs after tax credits.**

The payback period realized from savings after federal and state tax credits can be seen in Figure 8.4 to have been reduced to the following number of years for each example:

Example 1: From 16 years, 6 months to 10 years, 5 months (reduction of 6 years, 1 month)
Example 2: From 14 years, 11 months to 9 years, 2 months (reduction of 5 years, 9 months)
Example 3: From 16 years, 3 months to 10 years, 2 months (reduction of 6 years, 1 month)

If future worth of savings (as discussed in Chapter 6, Section 6.1) is considered throughout a 20-year period for each of the yearly savings in Table 8.3, column 5 at a 5% compounded energy inflation rate, the resultant future worth savings can be determined using Chapter 6, Eqn (6.2) and the single payment compound amount factors (SPCAF) in Table 6.1 for each year as follows:

Where:

S = future worth of money,
P = savings for a particular year (Table 8.3, column 5), and

(5%—SPCAF)=single payment compound amount factor (Table 6.1 factor)

$S_{20} = P$ (5%—20 SPCAF) = 845.22 (2.6533) = $2242.62
$S_{19} = P$ (5%—19 SPCAF) = 878.52 (2.5269) = $2219.93
$S_{18} = P$ (5%—18 SPCAF) = 913.03 (2.4066) = $2197.30
$S_{17} = P$ (5%—17 SPCAF) = 948.80 (2.2920) = $2174.65
$\vdots$ $\vdots$ $\vdots$ $\vdots$
$S_1 = P$ (5%—1 SPCAF) = 1725.93 (1.0500) = $1812.23

Table 8.4 Example of Utility-Provided Electricity Costs and Future Worth of Savings Realized

Years	Cost of Utility-Provided Electricity at 5% per year Energy Inflation		Savings Realized from Solar PV versus Electric Utility at 5% per year Energy Inflation		Future Worth of Savings through 20 years at 5% Compounded Inflation	
	(1) Yearly	(2) Cumulative	(3) Yearly	(4) Cumulative	(5) Yearly	(6) Cumulative
1	$853.76	$853.76	$845.22	$845.22	$2242.62	$2242.62
2	$896.45	$1750.21	$878.52	$1723.74	$2219.93	$4462.55
3	$941.27	$2691.48	$913.03	$2636.77	$2197.30	$6659.85
4	$988.33	$3679.81	$948.80	$3585.57	$2174.65	$8834.50
5	$1037.75	$4717.56	$985.86	$4571.43	$2152.03	$10,986.53
6	$1089.64	$5807.20	$1024.26	$5595.69	$2129.33	$13,115.86
7	$1144.12	$6951.32	$1064.03	$6659.72	$2106.67	$15,222.53
8	$1201.33	$8152.65	$1105.22	$7764.94	$2084.00	$17,306.53
9	$1261.39	$9414.04	$1147.86	$8912.80	$2061.44	$19,367.97
10	$1324.46	$10,738.50	$1192.01	$10,104.81	$2038.69	$21,406.66
11	$1390.69	$12,129.19	$1237.71	$11,342.52	$2016.11	$23,422.77
12	$1460.22	$13,589.41	$1284.99	$12,627.51	$1993.40	$25,416.17
13	$1533.23	$15,122.64	$1333.91	$13,961.42	$1970.85	$27,387.02
14	$1609.89	$16,732.53	$1384.51	$15,345.93	$1948.14	$29,335.16
15	$1690.39	$18,422.91	$1436.83	$16,782.75	$1925.50	$31,260.66
16	$1774.91	$20,197.82	$1490.92	$18,273.67	$1902.86	$33,163.52
17	$1863.65	$22,061.47	$1546.83	$19,820.50	$1880.17	$35,043.69
18	$1956.83	$24,018.30	$1604.60	$21,425.10	$1857.48	$36,901.17
19	$2054.68	$26,072.98	$1664.29	$23,089.39	$1834.88	$38,736.05
20	$2157.41	$28,230.39	$1725.93	$24,815.32	$1812.23	$40,548.28
	Total	$28,230.39	Total	$24,815.32	Total	$40,548.28

Developing the future worth of savings from the information previously included in Table 8.3, we have the yearly future worth of savings as shown in Table 8.4.

At this point, we should determine whether it would be wiser to invest in a solar PV energy system or in a bank savings account. So what is our money worth if we simply put the initial cost of the solar PV system into a savings account? Currently, we are only going to get a taxable 1% interest rate at best and more likely only 0.5% interest for our money in a bank (for rates available in 2012–2013). But, let's calculate which would be the better investment for each of our inverter combination examples even at a 1% rate of return from a savings account.

Example 1: Configuration with Microinverters

Future worth of $11,308.40 savings (amount banked versus expenditure for a PV system) for 20 years @ 1% compounded interest,

	Bank (Savings)
$S_{20}=P\,(1\%\text{—}20\ SPCAF) = (\$11,308.40) \times (1.220)$	$= +\$13,796.25$
Less expense of electricity costs from utility power company (over 20 years)	$= -\$28,230.39$
Actual money lost from bank savings of $11,308.40	$= -\mathbf{\$14,434.14}$
	Solar PV (Investment)
Future worth of energy savings for 20 years @ inflation rate of 5% compounded interest (Table 8.4, column 6) total over 20 years	$= +\$40,548.28$
Less expense for solar PV equipment and labor installation after federal and state tax credits	$= -\$11,308.40$
Actual money saved by installing a solar PV system	$= +\mathbf{\$29,239.88}$

Example 2: Configuration with One String Inverter

Future worth of $9625.60 savings (amount banked versus expenditure for a PV system) for 20 years @ 1% compounded interest,

	Bank (Savings)
$S_{20} = P\,(1\%\text{—}20\ SPCAF) = (\$9625.60) \times (1.220)$	$= +\$11,743.32$
Less expense of electricity costs from utility power company (over 20 years)	$= -\$28,230.39$
Actual money lost from bank savings of $11,308.40	$= -\mathbf{\$16,487.07}$
	Solar PV (Investment)
Future worth of energy savings for 20 years @ inflation rate of 5% compounded interest (Table 8.4, column 6) total over 20 years	$= +\$40,548.28$
Less expense for solar PV equipment and labor installation after federal and state tax credits	$= -\$9625.60$
Actual money saved by installing a solar PV system	$= +\mathbf{\$30,922.68}$

Example 3: Configuration with Two String Inverters

Future worth of $11,015.80 savings (amount banked versus expenditure for a PV system) for 20 years @ 1% compounded interest,

	Bank (Savings)
$S_{20} = P\,(1\%\text{—}20\ SPCAF) = (\$11,015.80) \times (1.220)$	$= +\$13,439.28$

Less expense of electricity costs from utility power company (over 20 years)	= −$28,230.39
Actual money lost from bank savings of $11,308.40	= **−$14,791.11**
	Solar PV (Investment)
Future worth of energy savings for 20 years @ inflation rate of 5% compounded interest (Table 8.4, column 6) total over 20 years	= +$40,548.28
Less expense for solar PV equipment and labor installation after federal and state tax credits	= −$11,015.80
Actual money saved by installing a solar PV system	= **+$29,532.48**

The savings bank investment in comparison with the three examples of solar PV investment clearly illustrates that the solar investment in supplementing electrical energy demands provides a positive cash savings. The solar investment also provides a better use of *cash flow* throughout the 20-year period. Considering the future worth of money over 20 years, Example 1 illustrates that purchasing a solar PV system produces a future worth difference of $43,674.02 between the actual money lost from a bank savings investment ($14,434.14) and the actual money saved ($29,239.88) from energy generated. Example 2 produces a $47,409.75 difference and Example 3 produces a $44,323.59 difference. The solar PV system cost, just like the solar DHW cost, is "self-liquidating," in that once the system cost has been repaid, there are few or no additional costs associated with the remaining years of normal system operation.

Although every solar PV system represents an individual case, an economical evaluation will indicate that solar energy is an excellent application in using the sun's energy to supply electricity rather than paying escalating electricity costs. The example costs and calculations used throughout this chapter have been conservative. It is assumed that the system is exempt from sales tax and exempt from property tax, which is true in some, but not all states. In many situations (i.e., using $0.16/kWh electrical rates), the return on investment averages less than 10 years.

Chapter nine

The Solar Investment

In July 1979, the State of Maine Office of Energy Resources reported the cost of electricity in that northern state to be $0.04/kWh and the cost of oil to be $0.57/gallon. By 2012, those costs were up 400 and 650%, respectively. The technology and practical economics of solar domestic hot water (DHW) systems were viable in the late 1970s, but photovoltaic (PV) systems were in the early stages of development and somewhat cost prohibitive. Since that time, the improvement in materials, increased efficiencies, and technological advancements have made both solar DHW systems *and* PV systems cost competitive in terms of economic payback periods versus fossil fuels. The economic climate for such systems has improved as predicted since that time because of escalating fuel costs, and the fact that tax incentives have been reintroduced, decreasing breakeven cost periods. Since the costs of fuel sources, unlike the sun's free radiant energy, continue to increase, it is important to understand the potential economic benefits of these alternative energy-producing systems. The economic portions of this book, therefore, are intended to provide a basic understanding of the economic decision-making process using examples throughout, but they are not intended as a complete course in managerial and financial engineering analysis.

The cost of installing a solar DHW system is an alternative to reducing the cost of using existing conventional energy systems, such as an electric, natural gas, oil burner, or other types of hot water delivery systems. Its purpose is to supplement the energy required to heat water for your residence. The cost of installing a PV system is an alternative supplement to reducing the cost of electrical energy received from your power utility company. So why would you pay more for equipment to supply energy for your daily requirements? The answer is fairly simple. The cost of energy from the sun remains free. Inflation does not affect its energy output. To measure the benefits of using these alternative energy systems, you need to add up the dollar savings from energy produced by each of these solar alternatives over their expected lifetime and compare the cost benefits against the actual costs. Economists call this type of analysis "life-cycle cost analysis".

This book addresses both payback periods and yearly cash flow over the expected life of a solar energy system to determine the value of a potential investment. If you have the ability to purchase a system with cash, you are looking for an investment comparison with that of other alternative investments. The family with the financial resources to purchase a system with cash has the most opportune position from an investment standpoint. If that is the case, the payback period and additional resulting return of equity over the life of the system will provide useful information regarding the actual value of the alternative energy system. The investment discussions in Chapters 7 and 8 were predicated on these types of evaluations. If you require financing a system with a loan, however, it is not necessary to compare the solar cost investment with any other alternative investment. If you finance a system, you are likely to be concerned with minimizing monthly energy expenditures. In that situation, a yearly cash-flow evaluation will illustrate the amount of your yearly loan payment that would be offset by the cost savings of your fuel or electric bill.

9.1 CASH-FLOW EVALUATION

An informative way to understand and examine a cash-flow analysis is to follow a particular example. Once the logic and calculation methods are understood, you then can substitute your own estimate costs relative to the type of alternative energy system under consideration and determine your own economic analysis.

9.1.1 Solar Domestic Hot Water System

Suppose you purchase the solar DHW system proposed in **Chapter 7, Section 7.6**, in the amount of $10,891 before tax credits. The actual system costs were as follows:

$$\begin{aligned}
\text{Solar DHW system cost} &= +\$10,891.00 \\
\text{Federal tax credit (30\%)} &= \underline{-\$\ \ \ 3267.30} \\
\text{Actual system cost after fed.tax credit} &= \$\ \ 7,623.70 \\
\text{State tax credit (i.e. billings, MT)} &= \underline{-\$\ \ \ 1000.00} \\
\text{Final system cost after tax credits} &= \$\ \ 6,623.70
\end{aligned}$$

Assume you have enough cash to pay the difference of $4267.30, which is the combined dollar amount of the federal ($3267.30) and state ($1000) tax credits allowed, so that at the end of the tax year, you are borrowing only the cost of the system, after tax credits, in the amount of $6623.70. From that previously proposed example, the output from the solar energy system provides 14.19 million British thermal units (MBTUs) per year. Per the example of Chapter 7, Section 7.5, we are heating water with electricity at $0.16/kWh at a cost of $665.22 for the first year. Table 9.1 illustrates the electrical energy cost with a 5% energy inflation factor. The cash needed to purchase 14.19 MBTUs of electricity at $0.16/kWh over a 10-year period, based on 5% energy inflation was calculated to be $8367.07. Average monthly payments for that amount of energy would be approximately $69.73 [$8367.07 (over 10 years)/120=$69.73].

Assume you borrow $6623.70 ($P$) to pay for the remaining portion of the system after receiving your tax rebates, and assume a bank loan for the system is obtained at 3.5% interest (i) compounded annually for a 10-year (n) term. The

Table 9.1 Example of Energy Costs and Savings Realized (System Output of 14.19 MBTUs at $0.16/kWh)

| Year | Conventional Electrical DHW Energy Costs at 5% per year Inflation Based on $0.16/kWh for 14.19 MBTUs (reference Table 7.5 of Section 7.5) | |
	Yearly	Cumulative
1	$665.22	$665.22
2	698.48	1363.70
3	733.41	2097.11
4	770.08	2867.18
5	808.58	3675.76
6	849.01	4524.77
7	891.46	5416.23
8	936.03	6352.26
9	982.83	7335.09
10	1031.97	8367.07
11	1083.57	9450.64
12	1137.75	10,588.39
13	1194.64	11,783.03
14	1254.37	13,037.40
15	1317.09	14,354.49
16	1382.94	15,737.44
17	1452.09	17,189.53
18	1524.70	18,714.22
19	1600.93	20,315.16
20	1680.98	21,996.13

series of loan repayments for each year can be found using Chapter 6, Eqn (6.5) and the applicable capital recovery factor (CRF), obtained from Table 6.1 under the applicable interest rate, as follows:

$$R = P\,(i - n\,\text{CRF})$$

Where
   R = repayment made at the end of each year,
   P = $6623.70 (present sum),
   $(i - n\,\text{CRF}) = (0.035 - 10\,\text{CRF}) = 0.12024$ (from Chapter 6, Table 6.1), and
   $R = (\$6623.70) \times (0.12024) = \mathbf{\$796.43}$ (annual payments for 10 years).

The series of payments (R) means that you would pay monthly installments of **$66.37** during the 10-year period to repay the loan. Table 9.2 shows that the amount of money borrowed at the start of the year (column 1) has interest due at the end of the year (column 2), and the payment at the end of the period (column 4) repays this interest plus some of the principal (column 6). For example, the unpaid principal at the beginning of year 3 is $5474.74, the interest owed that year at 3.5% is $191.62, and the payment at the end of the year of $796.43, consists of $191.62 in interest and $604.81 in principal.

Table 9.2 Visualizing the Capital Recovery Factor

Year	(1) Money Owed at Start of Year	(2) Interest Owed at End of Year at 3.5%	(3) (1)+(2) Principal and Interest Owed at End of Year	(4) Series (R) of Repayments	(5) (3)-(4) Money Owed at End of Year After Repayment	(6) (4)-(2) Recovery Capital
1	$6623.70	$231.83	$6855.53	$796.43	$6059.10	$564.60
2	$6059.10	$212.07	$6271.17	$796.43	$5474.74	$584.36
3	$5474.74	$191.62	$5666.36	$796.43	$4869.93	$604.81
4	$4869.93	$170.45	$5040.38	$796.43	$4243.95	$625.98
5	$4243.95	$148.54	$4392.49	$796.43	$3596.06	$647.89
6	$3596.06	$125.86	$3721.92	$796.43	$2925.49	$670.57
7	$2925.49	$102.39	$3027.88	$796.43	$2231.45	$694.04
8	$2231.45	$78.10	$2309.55	$796.43	$1513.12	$718.33
9	$1513.12	$52.96	$1566.08	$796.43	$769.65	$743.47
10	$769.65	$26.94	$796.59	$796.43	$0[1]	$769.49
					Total	$6623.54[1]

[1]Actual value calculates to $0.16 due to rounding to the nearest cent.

The actual cost of the solar DHW system in terms of present worth money at 3.5% compounded annually is the sum of the initial amount of money borrowed plus the present worth of the interest due at the end of each repayment period. In this case, using Chapter 6, Eqn (6.4) and Table 6.1, you can determine the actual cost of the system at 3.5% over a period of 10 years, as follows:

$$P = S\,(i - n \text{ SPPWF})$$

Where
P = present worth of interest due,
S = amount of the interest due = $231.83 (year 1) (column 2, Table 9.2),
SPPWF = single payment present worth factor,
$(i - n \text{ SPPWF}) = (0.035 - n \text{ SPPWF})$, resulting in the following tabulation:

$P_1 = S\,(3.5\% - 1\;\text{SPPWF}) = \$231.83\;(0.9662) = \$233.99$	
$P_2 = S\,(3.5\% - 2\;\text{SPPWF}) = \$212.07\;(0.9335) = \$197.97$	
$P_3 = S\,(3.5\% - 3\;\text{SPPWF}) = \$191.62\;(0.9019) = \$172.82$	
$P_4 = S\,(3.5\% - 4\;\text{SPPWF}) = \$170.45\;(0.8714) = \$148.53$	
$P_5 = S\,(3.5\% - 5\;\text{SPPWF}) = \$148.54\;(0.8420) = \$125.07$	
$P_6 = S\,(3.5\% - 6\;\text{SPPWF}) = \$125.86\;(0.8135) = \$102.39$	
$P_7 = S\,(3.5\% - 7\;\text{SPPWF}) = \$102.39\;(0.7860) = \$80.48$	
$P_8 = S\,(3.5\% - 8\;\text{SPPWF}) = \$78.10\;(0.7594) = \$59.31$	
$P_9 = S\,(3.5\% - 9\;\text{SPPWF}) = \$52.96\;(0.7337) = \$38.86$	
$P_{10} = S\,(3.5\% - 10\;\text{SPPWF}) = \$26.94\;(0.7089) = \$19.10.$	
Original system loan	$6623.70
Present worth system cost	**$7802.22**

In a loan repayment scenario of $6623.70 at 3.5% compounded interest with end-of-period repayments, the present worth of money actually invested, taking into account the present worth of the interest paid each year, would be **$7802.22**. The additional $1178.52 for the cost of the system (i.e., $7802.22 – $6623.70) essentially would be paid for by the cost savings of the energy produced. In approximately 9.5 years, the solar DHW system would begin to accumulate savings and decrease your monthly energy costs to heat water. In addition, during the 10-year loan period, the amount of money you pay for the energy produced would remain the same because the sun's energy is not affected by energy inflation costs, fuel availability, or politics.

Because the payments for the solar DHW system are **$66.37** each month, based on the series of repayments (R) calculated, you conservatively would be saving a few dollars over that period of time as well as pay for the cost of the system. The energy savings attributed to the operation of the solar installation would offset the interest paid on the loan. After that period of time, you would have a net positive savings of cash flow each month. Essentially, you have lost no more money than you otherwise would have spent, simply by paying the same monthly amount for your current hot water energy needs.

9.1.2 Solar Photovoltaic System

Calculating savings for electricity is fairly straightforward because efficiencies are calculated at 100%, and you are comparing costs using the same form of energy. Payback evaluations for PVs therefore are more definitive because you are comparing the cost of electricity production from a solar PV array with the cost of electricity from a power company. Analogous to the solar DHW system just discussed, suppose you purchase the PV system proposed in Chapter 8, Section 8.3, Example 1, in the amount of

$19,012 before tax credits. The actual system costs after tax credits were as follows:

$$
\begin{aligned}
(\textbf{Cost of system installed}) &= \quad \textbf{\$19,012.00} \\
\text{Federal tax credit (30\,\%)} &= -\$ \quad 5703.60
\end{aligned}
$$

$$
\begin{aligned}
\text{Actual system cost after fed.tax credit} &= \quad \$13,308.40 \\
\text{State tax credit} &= -\$ \quad 2000.00
\end{aligned}
$$

$$
\textbf{System cost after tax credits} = \quad \textbf{\$11,308.40}
$$

Assume you have enough cash to the pay the difference of $7703.60, which is the combined dollar amount of the federal ($5703.60) and state ($2000.00) tax credits allowed. At the end of the tax year, you are borrowing the actual cost of Example 1, after you receive your tax credits, in the amount of $11,308.40. From that previously proposed example, you are producing 5336 kW at a value of $0.16/kWh for the first year. Table 9.3 illustrates the electrical energy cost with a 5% energy inflation factor as represented previously in Chapter 8, Section 8.3, Table 8.3. The cash flow to purchase 5336 kW of electricity at $0.16/kWh over a 10-year period, based on 5% energy inflation was calculated to be $10,738.50. Average monthly payments for that amount of electricity would be approximately **$89.49**.

Table 9.3 Example of Annual Electrical Energy Costs (Electricity Demand of 5336 Kwh at $0.16/kWh)

Year	Conventional Utility Electrical Demand Costs at 5% per year Energy Inflation	
	Yearly	Cumulative
1	$853.76	$853.76
2	$896.45	$1750.21
3	$941.27	$2691.48
4	$988.33	$3679.81
5	$1037.75	$4717.56
6	$1089.64	$5807.20
7	$1144.12	$6951.32
8	$1201.33	$8152.65
9	$1261.39	$9414.04
10	$1324.46	$10738.50
11	$1390.69	$12129.19
12	$1460.22	$13589.41

Table 9.3 Example of Annual Electrical Energy Costs (Electricity Demand of 5336 Kwh at $0.16/kWh)

| Year | Conventional Utility Electrical Demand Costs at 5% per year Energy Inflation | |
	Yearly	Cumulative
13	$1533.23	$15122.64
14	$1609.89	$16732.53
15	$1690.39	$18422.91
16	$1774.91	$20197.82
17	$1863.65	$22061.47
18	$1956.83	$24018.30
19	$2054.68	$26072.98
20	$2157.41	$28230.39

Let's perform the same type of analysis for this PV system as we did previously in Section 9.1.1 for a DHW system. Assume you borrow $11,308.40 ($P$) to pay for the remaining portion of the system after receiving your tax rebates, and assume a bank loan for the system is obtained at 3.5% interest (i) compounded annually for a 10-year (n) term. The series of loan repayments for each year can be found using Chapter 6, Eqn (6.5) and the applicable capital recovery factor (CRF), obtained from Table 6.1 under the applicable interest rate, as follows:

$$R = P\,(i - n\,\text{CRF})$$

Where

R = repayment made at the end of each year,
P = **$11,308.40** (present sum),
$(i - n\,\text{CRF}) = (0.035 - 10\,\text{CRF}) = 0.12024$ (from Chapter 6, Table 6.1), and
$R = (\$11,308.40) \times (0.12024) = \textbf{\$1359.72}$ (annual payments for 10 years).

The series of payments (R) means that you would pay monthly installments of **$113.31** during the 10-year period to repay the loan. Table 9.4 shows that the amount of money borrowed at the start of the year (column 1) has interest due at the end of the year (column 2), and the payment at the end of the period (column 4) repays this interest plus some of the principal (column 6).

The actual cost of the PV system in terms of present worth money at 3.5% compounded annually is the sum of the initial amount of money borrowed plus the present worth of interest due at the end of each repayment period. Using Chapter 6, Eqn (6.4) and Table 6.1, you can

Table 9.4 Visualizing the Capital Recovery Factor

Year	(1) Money Owed at Start of Year	(2) Interest Owed at End of Year at 3.5%	(3) (1)+(2) Principal and Interest Owed at End of Year	(4) Series (R) of Repayments	(5) (3)-(4) Money Owed at End of Year after Repayment	(6) (4)-(2) Recovery Capital
1	$11,308.40	$395.79	$11,704.19	$1359.72	$10,344.47	$963.93
2	$10,344.47	$362.06	$10,706.53	$1359.72	$9346.81	$997.66
3	$9346.81	$327.14	$9673.95	$1359.72	$8314.23	$1032.58
4	$8314.23	$291.00	$8605.23	$1359.72	$7245.51	$1068.72
5	$7245.51	$253.59	$7499.10	$1359.72	$6139.38	$1106.13
6	$6139.38	$214.88	$6354.26	$1359.72	$4994.54	$1144.84
7	$4994.54	$174.81	$5169.35	$1359.72	$3809.63	$1184.91
8	$3809.63	$133.34	$3942.97	$1359.72	$2583.25	$1226.38
9	$2583.25	$90.41	$2673.66	$1359.72	$1313.94	$1269.31
10	$1313.94	445.99	$1359.93	$1359.72	0[1]	$1313.73
					Total	$11,308.19[1]

[1]Actual value calculates to $0.21 due to rounding to the nearest cent.

determine the actual cost of the system at 3.5% over a period of 10 years, as follows:

$$P = S\,(i - n\ SPPWF)$$

Where

P = present worth of interest due,
S = amount of the interest due = $395.79 (Year 1) (column 2, Table 9.4),
$(i - n\ SPPWF) = (0.035 - n\ SPPWF)$, resulting in the following tabulation:

$P_1 = S\,(3.5\% - 1\ SPPWF) = \$395.79\ (0.9662) = \$382.41$
$P_2 = S\,(3.5\% - 2\ SPPWF) = \$362.06\ (0.9335) = \$337.98$
$P_3 = S\,(3.5\% - 3\ SPPWF) = \$327.14\ (0.9019) = \$295.05$
$P_4 = S\,(3.5\% - 4\ SPPWF) = \$291.00\ (0.8714) = \$253.58$
$P_5 = S\,(3.5\% - 5\ SPPWF) = \$253.59\ (0.8420) = \$213.52$
$P_6 = S\,(3.5\% - 6\ SPPWF) = \$214.88\ (0.8135) = \$174.80$
$P_7 = S\,(3.5\% - 7\ SPPWF) = \$174.81\ (0.7860) = \$137.40$
$P_8 = S\,(3.5\% - 8\ SPPWF) = \$133.34\ (0.7594) = \$101.26$
$P_9 = S\,(3.5\% - 9\ SPPWF) = \$90.41\ (0.7337) = \$66.33$

$P_{10} = S\,(3.5\% - 10\;SPPWF) = \$45.99\,(0.7089) = \$32.60.$	
Original system loan	$11,308.40
Present worth system cost	**$13,303.33**

In a loan repayment scenario of $11,308.40 at 3.5% compounded inter-est with end-of-period repayments, the present worth of money actually invested, taking into account the present worth of the interest paid each year, would be **$13,303.33**. The additional $1994.93 for the cost of the system (i.e., $13,303.33 – $11,308.40) essentially would be paid for by the cost savings of the energy produced. In approximately 12 years, the PV system would begin to provide accumulative savings and decrease your monthly electricity bills.

Since the payments for the solar PV system are **$113.31** each month, based on the series of annual repayments (R) calculated for this scenario, you would be paying slightly more per month ($113.31 – $89.49 = $23.82) over this 10-year period to pay off the loan based on the conservative energy inflation percentages utilized. These monthly payments can be fine-tuned by your banking lender to actually result in the same amount of money spent for energy each month as the repayment of the loan. As discussed in Chapter 8, the breakeven and payback over the life of this system will produce a positive cash flow. By considering the energy output over the lifetime of a solar energy system, you can determine whether the overall benefits exceed the costs of a conventional system. This type of analysis has been presented in Chapter 7 by determining the payback or breakeven period discussed in regard to solar DHW systems and in Chapter 8 in regard to PV systems.

9.2 ENERGY CHOICES

Water heating accounts for approximately 20% of all household energy use in the United States,[1] and the use of solar DHW has the potential to reduce that household energy consumption by 50% or more as illustrated by the example presented in Chapter 7. Approximately 40% of the residential water heaters are electric[2] and 54% are natural gas.[3] The remainder of water is heated with either propane or oil-fired boilers, mostly in the northeast regions, where a quarter to a third of the residences use fuel oil.

Heating water is a function of the climate, inlet water temperature, and demand patterns of the consumer. The Northeast and Rocky Mountain regions generally have a higher water heating energy demand due to cooler inlet water temperatures, whereas the energy needed to heat water in warmer climates is

---

[1] Energy Center of Wisconsin; ECW Report 254-1; American Council for an Energy-Efficient Economy. NREL/TP-6A20-48986, February 2011. http://www.aceee.org.
[2] American Council for an Energy-Efficient Economy (www.aceee.org).
[3] National Renewable Energy Laboratory; Cassard, H, Denholm, P, Ong, S, Break-Even Cost for Residential Solar Water Heating in the United States: Key Drivers and Sensitivities. NREL/TP-6A20-48986; February 2011.

significantly less. Seasonal variations in solar insolation availability including precipitation, cloud cover, and ambient temperatures all play a part in the amount of radiant energy available. Such factors used to determine heating demands have been discussed in Chapter 3. System prices, quality, durability, fuel costs, utility rates, and tax incentives are just a few of the additional factors to consider. Although a conventional DWH system is less costly initially, it will cost more to heat the water over the life of the system than the cost of a solar energy system because of inflation and increasing fossil fuel costs. Comparing payback periods for solar DHW is more complicated than determining payback periods for PV systems, particularly because of inefficiencies with fuel to heat conversion. Calculating savings for oil, natural gas, and propane across different system technologies and types of configurations is a bit more difficult.

9.2.1 Fuel Oil

Let's take a look at the cost of fuel for an oil burner to provide the same amount of energy to heat water as electricity. Table 9.5 is similar to Chapter 7, Section 7.1, Table 7.3 except we are now using the cost of oil versus electricity. Assuming an 85% efficiency, a gallon of fuel oil provides 117,895 BTUs of energy (see Chapter 3, Section 3.1, Table 3.1). The number of gallons of oil necessary to produce 17.3 MBTUs ((17,300,000 BTUs)×(1 gallon/117,895 BTUs)) is 146.7 gallons. The number of gallons used, shown in Table 9.5, is based on the efficiency of the oil burner and can be calculated accordingly for each individual situation.

At first glance, it would appear that it would be less expensive to heat water with oil than with electricity. In 2012, the Northeast region of the country was paying approximately $0.16/kWhKwh for electricity and $3.70/gallon for oil. At those prices, Table 7.3 shows the cost to heat 60 gallons of water by electricity to be $810.78, compared with $542.75 for oil (Table 9.5). The results of this comparison can be somewhat deceiving, however, because they do not take into consideration the inconsistencies in oil burner system efficiencies and heat losses for the warmer months just to generate hot water. "During the summer months, boilers without cold start functionality can have efficiencies approaching 25% with indirect tanks and tankless coils alike. It is not uncommon to consume between 100 and 200 gallons of oil for two occupants over the summer months for water heating."[4] So how does this affect using oil versus electricity in comparison to using solar?

If the price of oil is $3.70/gallon, and an additional amount of oil consumed during the seasonal warmer months is conservatively 100 gallons due to heat losses, then you have added an additional $370 to your fuel costs in Table 9.5. This results in an annual cost of $912.75 ($542.75+$370), which is approximately $102 more than the cost of heating water with electricity. If you apply these costs to a solar DHW system output of 14.19 MBTUs as discussed previously in

---

[4] Fischer, D., May 7, 2012. Efficiency Maine. Analysis comment with permission.

Table 9.5 Typical Domestic Hot Water Provided by Fuel Oil

Water Heated per day (gallon)	Yearly Requirement MBTU	Gallons of Oil	Yearly Cost to Heat Water from 40 to 135°F $3.50/gallon	$3.70/gallon	$3.90/gallon	$4.10/gallon	$4.30/gallon	$4.50/gallon	$4.70/gallon	$4.90/gallon
60	17.3	146.7	$513.45	$542.75	$572.13	$601.47	$630.81	$660.15	$689.49	$718.83
70	20.2	171.3	$599.55	$633.81	$668.07	$702.33	$736.55	$770.85	$805.11	$839.37
80	23.1	195.9	$685.65	$724.83	$764.01	$803.19	$842.37	$881.55	$920.73	$959.91
90	26.0	220.5	$771.75	$815.85	$859.95	$904.05	$948.15	$992.25	$1036.35	$1080.45
100	28.9	245.1	$857.85	$906.87	$955.85	$1004.91	$1053.93	$1102.50	$1151.97	$1200.99
110	31.8	269.7	$943.95	$997.85	$1051.83	$1105.77	$1159.71	$1213.65	$1267.59	$1321.53
120	34.7	294.3	$1030.05	$1088.91	$1147.77	$1206.63	$1265.49	$1324.35	$1383.21	$1442.07

Chapter 7, Table 7.5, for the family of four in Billings, Montana, you would use 120.4 gallons of oil ((14,190,000 BTUs) × (1 gallon/117,895 BTUs) = 120.4 gallons) plus 100 gallons for heating water in the summer months for a total of 220.4 gallons annually. At $3.70/gallon, the cost of fuel oil is $815.48, which is $150.26 ($815.48 − $665.22) more than the cost of heating with electricity, shown in Chapter 7, Table 7.5.

In the northern states, oil consumption use can be diminished by installing a solar DHW system and by converting the boiler to a cold-start operational mode. Using this combination, the boiler remains off unless the thermostat calls for heat when there is not sufficient sun for backup heat in the solar tank. This combination of solar DHW and digital aquastat for the boiler will result in a shorter payback period than that shown in Chapter 7, Figure 7.3.

9.2.2 Natural Gas

Natural gas is lower in cost per BTU than either electricity or oil, and therefore, the breakeven cost of natural gas versus solar DHW for the same time period is higher. In a study performed by the U.S. Department of Energy in February 2011,[3] it was noted that the cost of a solar DHW system would have to be reduced by 36% to break even with cost of using natural gas. Does this mean that if you heat water using natural gas that you should not look into using solar to supplement the water heating process? If your current source for heating water is calculated to be less expensive than the cost of a solar DHW system and the overall savings and benefits do not exceed the actual costs of a system, then solar DHW may not be economically practical. This can be the situation when you compare natural gas prices against other fuel sources. When considering this type of investment, however, you must consider the volatility of future energy pricing.

The national annual average residential price of natural gas in 2009 and 2010 was $12.90/1000ft^3 ($12.90/mcf). From 2008 through 2012, the price of natural gas declined from a high peak of $20.77/mcf to an average peak of $15.85/mcf.[5] Using the energy conversion equivalencies as discussed in Chapter 3, Section 3.1, we can translate the dollar amount from cubic feet to therms, as follows:

$$(100,000 \text{ BTUs/therm}) \times (1 \text{ ft.}^3/1,028 \text{ BTUs})$$
$$\times (\$15.85/1000 \text{ ft.}^3) = \$1.54 \text{ /therm}$$

Therefore, if you used natural gas to produce 14.19 MBTUs per our previous example, it would take approximately 142 therms annually or only $219 to heat water. Natural gas prices, however, can be volatile as noted by a *Wall Street Journal* market watch report in June 2012, which announced that natural gas supplies had been reduced because of a 38% drop in the number of operational gas rigs, and predictions for the following year could be more than $3.00/therm.[5] In addition,

---

[5] U.S. Energy Information Administration. Independent Statistics and Analysis. Natural Gas. 6/2013. U.S. Natural Gas Price Graphical Data. http://www.eia.gov/forecasts.

natural gas prices can more than double from one state to another because of such factors as the number of pipelines in the state, the market's proximity to producing areas, transportation charges associated with delivery, average consumption per residential customer, state regulations, and degree of competition in the area.

9.2.3 Electricity

The cost of electricity in comparison with both solar DHW and PV systems was addressed in Chapters 7 and 8, respectively. In many scenarios, electricity demands from your local power utility company are used entirely for household needs, inclusive of DHW. You can evaluate a life-cycle cost analysis and determine the life-cycle benefits by following the examples provided. The comparison of electricity costs for your residential power demands, however, may provide a better breakeven cost analysis using a PV system rather than a solar DHW system. Both types of systems should be evaluated.

Life-cycle benefits often do not greatly exceed the capital cost of a system and benefits, such as reduced reliance on fossil fuels, are external to the consumer and difficult to quantify. Energy prices are changing constantly and an analysis to determine payback and long-term investment therefore are difficult to evaluate, but you can modify the tables within these chapters to include your own information to evaluate any type of cost scenario.

Because each homeowner has different patterns of energy usage and requirements as well as financial abilities, blank work tables are provided in Appendix B for use in developing your own financial analysis. Using the financial examples presented in this chapter and in Chapters 7 and 8, you can provide data particular to your situation and future changes in inflation factors to determine cost breakeven points, cash flow, and loan considerations.

9.3 ENERGY DECISIONS

This book started by discussing the reasons for considering all types of energy to achieve energy independence and use more alternative forms, in particular our sun. Increasing the availability and use of all domestic energy sources will lead to an overall improvement in our economy and economic stability. This is not a political statement. It is simply a fact. Using our own developed energy resources should allow us an increased opportunity to invest in alternative energy solutions. We should not be complacent and continue to rely on foreign fuel imports.

With the information in this book, you can determine whether or not you have the proper site to take advantage of the radiant energy available. You can determine energy equivalents and understand the methods behind heat transfer for DHW use as well as the knowledge needed to determine your electrical demands. You can determine the sizing of both solar DHW collectors and PV arrays, and you have a basic understanding of the components involved with both types of systems. In addition, you have the ability to compare the many manufacturer's models of solar DHW collectors and PV modules to determine the best fit for your particular application. Most important, you have an understanding of the economic payback and cash-flow analysis required to make an informed decision about whether to install a solar PV or DHW system.

Several websites have been mentioned throughout this book providing shortcuts and quick calculations for determining amounts of radiant energy, energy output, and collector performance. Website addresses can change over time and such online calculators can disappear. Manual tables have been included so that you do not have to depend on a computer to determine and evaluate these two alternative energy systems. A database of State Incentives for Renewables and Efficiency tax credits and rebates is also available at the U.S. Department of Energy Database of State Incentives for Renewables and Efficiency (www.dsireusa.org). Such up-to-date information is available via the Internet at that address or via your browser for other links to similar websites. Because time allowances and requirements for each state vary dramatically from one another, such information should be checked with each state before purchase. A qualified solar dealer or installer should have information regarding refunds, rebates, sales tax, and property tax exemptions, and building permit requirements applicable to your particular town and state.

Whether a PV system or solar DHW system is more cost-effective as an investment depends on several factors that vary from region to region and state to state, causing a variation in breakeven costs. In comparison with conventional systems, high initial costs are the primary reason for low consumer adoption. Solar energy installation prices can vary considerably from region to region, and the cost of comparative systems can vary significantly. Issues with aesthetics, system reliability, and lack of familiarity and knowledge about the technologies have combined to limit consumer adoption as well. Benefits such as reduced reliance on fossil fuels and reduced carbon-dioxide emissions can be difficult for consumers to quantify.

A solar energy system will increase the value of your property and can decrease your monthly energy expenditures, increasing your cash flow by saving money. This can be more advantageous than investing money in an alternative savings bank. Without the federal and state tax credits, however, the life-cycle benefits often do not greatly exceed the capital costs. With or without the tax credits, it is important to remember that there *is* a tax advantage with a solar alternative energy system investment. If you make money with a traditional investment, such as stocks or bonds, or even the low interest rates from a savings bank, your net income is increased and so is your tax liability. The money you invest in a solar DHW or PV system increases your spending power by saving you money. This is because you maintain the same income level when you save money, so your tax liability remains the same. You have not added to taxable income. You have simply spent less money; something the federal government should consider. Although alternative energy technologies continue to evolve, the financial relationships will remain the same. Only the cost of fuels, inflation rates, and loan interest rates are likely to rise. Everything else is subject to our perception of practicality and economic viability.

Chapter ten

The Energy Conundrum and Economic Consequences

"Wisdom will repudiate thee, if thou think to enquire Why things are as they are or whence they came; thy task is first to learn What Is…"

Robert Bridges

Not only is it important to understand the basic technical aspects and economic benefits of using solar energy alternative systems as discussed in the previous chapters, but it is also important to consider the possible consequences of mandating increased energy policies and regulations. The results of interventions and overburdening bureaucratic regulations actually can affect the growth of our economy, inhibiting the purchase of such systems and therefore curtailing their application. You should remember that there are always differing views and opinions to an issue, and it is important that we consider the implications of changes linked to energy policies. No matter what the issue, understand it and be careful what you wish for. This chapter addresses some of those concerns.

On a state-by-state level, the Edison Electric Institute stated in a September 2013 report that net metering was a threat to investor-owned utility companies in that net-metered customers were effectively avoiding paying grid-related costs, such as poles, meters, wires, and infrastructures. Even though most solar photovoltaic (PV) residential customers were purchasing additional power to supplement their additional energy requirements, it was suggested that policies and rate structures in many states should be updated so that everyone who uses the electric grid helps to pay to maintain it. As a result, several states—including California, Arizona, Georgia, and Colorado—have sought to add a surcharge or a monthly grid-maintenance charge as a regressive way for solar ratepayers to further support and pay for grid infrastructure. Such add-on fees, if not closely controlled, could reduce the long-term investment advantages of PV installations. It's important to be conscious that some investor-based utility companies may try to discourage net metering so you do not own your own power generating system. Such companies should recognize the need for renewable energy growth and should not be concerned only about corporate profits. They should recognize that distributed power generation paid for and contributed by individual investors save money on fuel as well as infrastructure, thus benefiting all ratepayers.

On a federal level, whatever happened to the 32 solar collectors that were installed at the White House as an experiment to provide hot water to the West Wing offices back in 1979? At that time, the Iranian revolution had thrown world oil markets into turmoil causing a U.S. energy crisis. Public information indicates that those solar panels were removed by aides to President Reagan because the administration was unimpressed with the system performance. The majority of the solar panels were later shipped to Unity College in the State of Maine where some were restored and remain in service. Since then, President George W. Bush's administration had a few modest PV systems installed on the roofs of several maintenance buildings to generate small amounts of power for the White House complex and to heat water for the mansion's pool. In 2010, the energy secretary said the administration would conduct a competitive-bidding process to purchase solar panels. Through 2012, however, no additional solar energy systems were established at the White House, even with the increased advocacy of "green energy". Installation of PV panels for the president's living quarters began in 2013.

The Department of Energy made purchasing solar energy systems for federal facilities more complicated when it published a 104-page handbook in September 2010 entitled, "Procuring Solar Energy: A Guide for Federal Decision Makers," mandating federal procurement requirements for solar energy. This handbook added complicated procedures and regulations for government agencies doing business with solar energy companies. The guidelines in this handbook begin by stating that "Because solar energy technologies are relatively new, their deployment poses unique challenges." Although recent improvements have been made, these technologies have been available for several decades; they are not really new. Over the past few years, many improvements *have been made* to the individual components that make up these systems, increasing their efficiency and ease of installation, but their overall designs, for the most part, have remained fundamentally consistent and dependable. Reading through the handbook could lead one to believe that the methods of procurement are *really* the new issue, not the solar energy technology. The executive summary notes that there is a two-part process needed to implement a smooth and successful solar project. It states, "Part 1 of the process includes five project planning steps that cover identification of needs and goals, assembling an on-site team, evaluating the site's solar screening, project requirements and recommendations, and making a financing and contracting decision." The summary continues stating that "Part 2 of the process includes process guidance on the following financing and contracting options: agency-funded projects, power purchase agreements, energy savings performance contracts, utility energy services contracts, and enhanced use leases." Reviewing all the in-depth regulations involved with these methods of procurement, included with this handbook, emphasizes the reasons why we cannot get much accomplished at the federal level.

Many federal "green energy" programs simply do not work very well. For example, the mandated use of corn by the Renewable Fuel Standards (RFS) program required at least 37% of the 2011–2012 corn crop be converted to ethanol and blended with gasoline to power our vehicles. The Congressional Budget Office indicated that this requirement ultimately raises the prices that consumers pay for a wide variety of foods at the grocery stores, ranging from corn syrup sweeteners to dairy and poultry products. This mandate to burn food in our automobiles places our corn food supply at increased risk of potential supply disruptions caused by drought and bad weather. The RFS mandate requires that a massive amount of corn be converted to ethanol each year regardless of price or available supply. Cars are therefore now at the top of our food chain. Has anyone ever had any issues with the quality of the gasoline being used with such additives, including lower fuel mileage, damage to fuel lines and other elastomers, and corrosion issues?

In many instances, Environmental Protection Agency (EPA) regulations are now so burdensome that they result in costly financial issues that ultimately are transferred to the consumer. For instance, the *New York Times*[1] published an article in 2012 reporting that refineries supplying motor fuel would pay the EPA a penalty of $6.8 million for not mixing in a percentage of cellulosic biofuel into their gasoline and diesel products for the tax year 2011. Unfortunately, this special biofuel did not exist except in a laboratory. Because of technological problems with developing this biofuel, it was not commercially available in 2011. Yet, the quota and penalties for not using this nonexistent biofuel actually increased. Apparently, the EPA set this quota for cellulosic ethanol while ignoring what the industry realistically could provide.

We all know that the cost of energy from fossil fuel production is volatile. For the most part, we can all understand there is a need to account for excessive regulations that may result from actions taken regarding climate change and greenhouse gas (GHG) emissions in economic terms. Historic research has shown that Greenland, located at its high latitude, has twice experienced dramatic and abrupt shifts in climate during the past 1000 years. The orbital dynamics of the earth, due to its precessional wobble, influences the variations in ice-age occurrences. As such, there is an intricate balance of nature's and potential human involvement with respect to climate change that has been under intense scrutiny and study for several years. As a result of such studies, research continues to evaluate the effect of our reliance on fossil-based fuels and the amount of carbon dioxide added to the atmosphere in the complex study of our planet's climate.

In March 2009, there was a 98-page draft report written by the Office of Policy, Economics, and Innovation Office of the Administrator U.S. Environmental Agency entitled "Proposed NCEE Comments on Draft Technical Support Document for Endangerment Analysis for Greenhouse Gas Emissions

---

[1] Wald, M.L. The New York Times. Business Day-Energy and Environment. A Fine for Not Using a Biofuel That Doesn't Exist. September 1, 2012.

under the Clean Air Act." This report apparently was not immediately released by the EPA, but rather it was withheld from publication. It stated that previous scientific findings were out of date, vehemently contradicting Section 701 under Title VII of the originally proposed Bill HR 2454 (American Clean Energy and Security Act). The more current data that are relevant appeared to greatly influence the assessment of "vulnerability, risk, and impacts" of climate change to the United States. Since the cut-off date of the United Nation's Intergovernmental Panel on Climate Change (IPCC) (AR4) Fourth Assessment Report, new developments were discussed in the unpublished report. This draft EPA report stated that "there are critical and disturbing inconsistencies in the data concerning the causes of global warming." In effect, global temperatures had declined according to the draft EPA report. It further stated that "it is not reasonable to conclude that there is any endangerment from changes in greenhouse levels based on the satellite record, since almost all the fluctuations appear to be due to natural causes and not human–caused pollution as defined by the Clean Air Act."

On June 26, 2009, EPA management gave permission to the author to post the report on a personal website, and on August 5, 2009, the EPA released the original March 16 version as a frequently requested record under the Freedom of Information Act. A nonsubstantively modified version of the March 16 version entitled "Comments on Draft Technical Support Document for Endangerment Analysis for Greenhouse Gas Emissions under the Clean Air Act" was prepared in late June 2009 by the author and highlights of that report are included as Appendix C. (Inconsistencies between the Technical Support Document analysis and the conclusions of the four individual IPCC (AR4) reports: the AR Synthesis Report; the Physical Science Basis Report; the Impact, Adaption, and Vulnerability Report; and the Mitigation of Climate Change Report are addressed in Appendix C.)

This issue of climate change was further expanded on in a Cap and Trade Bill that Congress attempted to pass. In 2009, Bill H.R. 2454 (the American Clean Energy and Security Act; Cap and Trade) was introduced to Congress embarking on a national campaign to federally control a "clean" energy vision. It consisted of more than 1400 pages that most people in Congress did not understand or likely even read. Ultimately, it did not pass. Deservedly so, however, there are concerns that this type of legislation will continue to be introduced in the future either at the federal level by Congress, on a regulatory level by the EPA, or on a global level by the United Nations.

Let's attempt to remove the prejudices from both the conservative as well as liberal sides and discuss cap and trade. Reducing carbon emissions through a cap-and-trade system means replacing the least expensive forms of energy (which include coal, oil, and natural gas) with currently more expensive sources, such as biofuels, wind, and solar. Wind and solar are good alternatives to supplement our energy base, but they cannot completely replace our basic energy structure in the near future. The carbon tax mandate is probably one of the more damaging bills introduced by Congress because of its effect on our economy. How many people really know what was in the proposed Cap and Trade Bill?

Let's elaborate on the concerns regarding such proposed mandates, just in case they happen to reappear in some format.

In accordance with Bill H.R. 2454, the term cap-and-trade means a system of GHG regulations under which a government entity issues a limited number of tradable instruments in the nature of emission allowances and requires that sources within its jurisdiction surrender such tradable instruments for each unit of GHG emitted during a compliance period. To force this conversion, Bill HR 2454 (with amendments) was introduced, bringing about crippling global warming economic regulations, essentially taxing the use of fossil fuels to make them prohibitively expensive. Levying such a tax through a cap-and-trade scheme would have had dire economic consequences on our country. Such an energy bill would have been one of the largest tax hikes in the country's history and ultimately would have caused more job losses. As if cap-and-trade was not sufficient in producing damaging consequences, the bill also included many other wasteful and costly programs, all of which added overwhelming government controls and regulations. The additional increase in the cost of energy and tax burdens caused by creating additional state and federal government agencies would have significantly reduced the amount of money available to the average taxpayer, preventing them from independently purchasing energy-saving-related products. In addition, utility rates ultimately would have increased significantly to all consumers due to cap-and-trade even with the use of emission allowances and the complex guidelines included in Sections 782–785 and throughout other portions of this bill. The amount of energy used ultimately would have been limited by what people could afford because utility rates would increase continually if emission limits were excessively overregulated. Such rate increases would severely affect the low- and middle-income families. Consumer product costs would increase, production would be reduced, and more unemployment would follow.

If Bill HR 2454 had become law, it may have created a few more green jobs (along with countless additions of government jobs to enforce and administer the bill), but it ultimately would have resulted in the loss of many more jobs than it created because of the dramatic increase in costs to all Americans. Most of the immediate green jobs created would have been state and federal government jobs. These jobs included many additional agencies and unnecessary boards, which would not have provided goods and services that expand our economy, but rather build on the infrastructure limiting our development and energy use. Program and agency types of jobs are not revenue producing but rather are tax burdening.

Let's briefly review several sections of the once-proposed 1427-page bill:

1. Section 114 appeared to regulate the quantity of fossil fuel for electricity delivered to retail customers by each distribution utility and included assessment fees of approximately $100 million annually. It's fairly obvious who ultimately would pay for these fees.
2. Section 131 established state-level managing and accounting of emission allowances, adding more administration regulations. Such policies provide our nation with wasteful monitoring and managing rather than providing actual constructive initiatives. It would be more useful to provide

information for the public regarding solar hot water, solar PV, and wind-energy systems. A less intrusive but more inclusive administration system could be established similar to the Northeast Solar Energy Center that was established in the early 1980s rather than that presented in Section 131. Installation guidelines for such systems similar to ones published in 1979 (i.e., HUD-PDR-407) by the U.S. Department of Housing and Urban Development in cooperation with the U.S. Department of Energy would be much more useful than adding layers of bureaucratic agencies for monitoring and creating laws resulting in wasteful spending.

3. Section 186 established additional bureaucracies with the formation of a separate corporation known as the Clean Energy Deployment Administration, with the ability to be totally independent of other agencies to "maximize the value of investments and promote clean energy technologies."

4. Section 304 established building code standards mandated by the federal government with federal funding penalties for noncompliance. Additional caveats allowed the secretary of energy to "set and collect reasonable inspection fees to cover costs of inspections required for such enforcement." It also included wording such that penalties would be applied to violators in any jurisdiction in which the national energy efficiency building code had been made applicable. Any states in noncompliance would be "ineligible to receive emission allowances (i.e., cap and trade again) and federal funding in excess of that state's share." Enforcement costs of this section were allocated at $25 million dollars annually. At this point, we start to see more of a loss of individual rights. Again, there appears to be too much government intervention within this original bill.

5. Subtitle I under the Bill was titled "Nuclear and Advanced Technologies", yet the sections defined under this title, including deployment goals in Section 185, provided no mention of nuclear or oil production programs. Only a report on the use of thorium-fueled reactors was requested by the International Atomic Energy Agency in Section 199A as a study. The word "nuclear" was mentioned only three times within the context of this 1427-page Energy Bill.

Construction of nuclear power plants would not only enhance our energy resource base, but also would provide short- and long-term jobs. Such technologically advanced sources of energy would greatly reduce our oil imports and at the same time drastically reduce carbon emissions. Drilling for oil off the U.S. coasts would add to that energy base, and a combination of energy alternatives including solar and wind would provide the United States with long-term economic and security benefits as well. Even the state of California would have been in less debt if drilling royalty revenue existed. It should be noted that drilling practices and technology have improved over the years, and it does not take a proverbial decade to get results. Only those politicians who do not want to address immediate drilling are convinced it takes 10 years or more to deliver oil from offshore wells. It all depends on the location, methods of drilling, and, of course, the permitting regulations. Offshore wells can take as little as 3 years depending on water depth and conditions. It also depends on how an oil company decides to approach the potential site by performing extensive surveys and exploration. Commenting solely on great lengths of initial production time provides an ongoing excuse for not drilling. There is the possibility that far more oil seeps naturally into our oceans from the ocean beds than from drilling. This so-called Energy Bill *never* addressed drilling for oil. In fact, the word "drilling" did not appear once in the 1427 pages. Again, these energy sources provide real revenue-producing jobs for the United States, not increased bureaucracy-enhanced government agencies.

6. Section 202 of Bill HR 2454 established the Retrofit for Energy and Environmental Performance program and had funding of $70 million annually for administration costs. It would have been far more effective to fund energy conservation efforts rather than to establish controls and accounting procedures.

7. Section 204 of Bill HR 2454 established a labeling program to initiate an efficiency rating of residential housing. Implementation included "disclosure of building label contents in tax, title, and other records" that localities would maintain and would be "adopted by statute or regulation that buildings be assessed and labeled" within each state. The preparation and disclosure of the efficiency label information would be gathered by several means, including "a final inspection of major renovations or additions made to a building in accordance with a building permit issued by a local government entity." Wording is ambiguous in one statement of this section that says, "no State shall implement a new labeling program pursuant to this section in a manner that requires the labeling of a building to occur after a contract has been executed for the sale of that building and before the sales transaction is completed." This means that all buildings would require an energy-efficiency rating before sales consideration. (This is not the only ambiguous wording in this bill.) Also included were another $60–$70 million annually to fund the administration of this program.

8. Section 213 suggested that appliances must meet specific energy conservation standards under applicable building codes. This section insinuates that an individual had to select particular models of an appliance, being ambiguous in its presentation.

9. Section 214 provided bonus payments to retailers for replacement of older operating low-efficiency products by the secretary of energy. This should be considered as excess spending by the government because under normal living conditions, people replace their appliances as needed. This should have been just business as usual. This section even stipulated that the secretary of energy would ensure that the older product is not returned to service, which means that large amounts of documentation would necessarily accompany these types of "awards". This section included bounty payments for refrigerants, bonus payments, premium award payments, and in-class incentives for appliances having lower energy consumption. There was "$600 million ($600,000,000) authorized to be appropriated for each of the fiscal years

2011 through 2013 to the Secretary of Energy for purposes of this section, and such sums as may be necessary for subsequent fiscal years." The costs of these incentives and administration would have weighed on the taxpayer and undoubtedly would have been more expensive than the cost of the energy saved.

10. Section 701 under Title VII—Global Warming Pollution Reduction Program concluded that Congress had found that "global warming poses a significant threat to national security, economy, public health and welfare and environment." There appears to be a disparity in such a statement because data are not conclusive as noted earlier. The issue remains debatable within the scientific community. To attempt to justify such a wasteful bill as HR 2454, it would be necessary to regard this issue as concluded. It is not, however, and the premise of this section is inconclusive, making a somewhat poor presumption without a confirmed basis.

11. Section 723 discussed the penalty for noncompliance for a covered entity to emit GHG and denoted attributable GHG emissions in excess of allowable emissions as defined within the bill. It was stated that each ton of carbon-dioxide equivalent that is noncompliant shall be a separate violation and that the amount of an excess emissions penalty shall be equal to the product obtained by multiplying the tons of carbon-dioxide equivalent of GHG emissions by "twice the auction clearing price for the earliest vintage year emission allowances in the last auction." So the direct costs of penalties at that point would appear to be indeterminate. This would have severely penalized the producers of electricity that use coal to keep costs low while trying to maintain standards.

12. Section 724 permitted lawful holders of an emission allowance, compensatory allowance, or offset credit the right to sell, transfer, and exchange them without restriction. There would be a system established for tracking, issuing, recording, and holding emission allowances, offset credits, and term offset credits. If not regulated properly (which implies more government monitoring), this could resemble another food-for-oil[2] debacle.

13. Section 743 discussed issuance of international offset credits for activities that take place outside of the United States, and Section 765 proposed establishing international "binding agreements" committing to reductions in industrial emissions. These sections required that the United States be a party to a bilateral or multilateral agreement with other countries. Such an arrangement would be problematic. This act of global governance could be viewed as an international global warming treaty (whether or not actual data supports the trend) that would be the first step toward international control of the U.S. energy supply, and ultimately the U.S. economy. There are horrific tax implications with such global governance, likely to include an international tax on carbon emissions where payments would be made

---

[2] Otterman, S. "Oil for Food Scandal". October 28, 2005.

to the United Nations. Such a proposal would cost U.S. taxpayers hundreds of millions of dollars, while potentially crippling our economy. This section alone implied long-term undefined consequences.

14. Section 793 established a strategic reserve fund, a climate change consumer refund account, and a climate change worker adjustment assistance fund. The reserve fund received proceeds from strategic reserve auctions, the climate change consumer refund account received proceeds from the auction of emission allowances after 2021, and the worker adjustment assistance fund received 0.5% of the emission allowances from 2012 through 2021 and 1% from 2022 through 2050. Administration costs to implement this overly complex bill would have been subject to costing $10 million for each fiscal year, further escalating the national deficit.

15. Sections 421–423 established a clean energy curriculum development program to aid career and technical education and job training for renewable energy sectors, which is laudable for supporting the creation of career jobs. Sections 425–427, however, established a program that entitled any worker displaced as a result of the Title VII Clean Air Act to be entitled for up to 3 years at up to 70% of the average weekly wage of such workers, 80% of their monthly health care premium, up to $1500 for job search assistance, and up to $1500 for moving assistance. Such entitlements are excessive and overwhelmingly exceed current federal unemployment benefits, causing unnecessary excessive increases to an uncontrollable national deficit. In addition, as commented on regarding Section 793, the operation and administration of this program would further escalate costs.

The preceding sections of the Cap and Trade Bill discussed are just a few of the fundamental observations made regarding the attempted bureaucratic increases in the number of oversight and assurance agencies, administrators, auditors, commissions, advisory boards, interagency groups, committees, and programs from being established. These prophetic increases are overwhelming, unnecessary, wasteful, and costly. Based on this bill, we could forget about energy independence because the cost of this bill would have prevented us from developing and implementing any of our own natural resources. The proposed Cap and Trade Bill would have rushed us to no solution at all but rather would have created an immense burden on the citizens of the United States while completely missing the goal of energy independence.

Why discuss a bill that never passed? Portions of this bill have been addressed in this book so that people are aware of the detrimental consequences that can evolve while attempting to control carbon emissions on a bureaucratic level. Many sources of data and observations in reports such as the excerpts provided in Appendix C indicate the climate-forcing effects of anthropogenic carbon dioxide are much less than the consensus of advocates of global warming. Scientific data appears to indicate the effects are small and that the trillions of dollars poised to spend on carbon-dioxide mitigation will have no effect, other than to slow the global economy and impede the goal of energy independence.

Climate is not only affected by carbon dioxide but by other gases as well, including nitrous oxide that can be 150 times more potent than GHG. This originates from the use of manure containing nitrogen and other fertilizers. It is produced with methane in wetlands as well as the oceans, complicating interpretations of the effects on climate.

Each type of energy production has its problematic adversities as well as merits. We worry about the safety of using nuclear reactors to generate electricity, but we have a substantially sound safety record for nuclear power in the United States. Even the use of natural gas can have safety issues as well. It also is easy to overlook the fact that even the use of oversized windmills are damaging in that they kill millions of birds, and bats have been found dead because of their echolocating. Because of the blade velocity of these "mills", their small organs literally explode inside of them as they plummet to their death. There is also a "flicker" effect from the blades, causing continuous shadows on trees and other structures. Solar is an alternative energy source that has fewer environmental repercussions than all other forms. In addition to economics, all energy sources need to be considered for their safety as well as their use and their environmental effects.

The development and use of alternative energies can supplement our energy needs, but they cannot completely satisfy them in the near term. As an example, waste wood in the form of wood pellets contributes to a cleaner energy supply, but its volume will not suffice to make a significant difference (with the possible exception of a few regional areas such as New England). Wood bonds carbon, and as long as it remains in such storage, it does not enter the atmosphere. Other carbon dioxide–free means include solar, windmills, nuclear energy, and hydroelectric. It is the inclusion of all forms of energy and conservation that will lead us to energy independence. Developing a mix of energy alternatives will bring about a more stable energy future; however, these alternatives do not necessarily have comparable capabilities. The illustration in Figure 10.1 compares annual energy consumption of the world to the known reserves of the *finite* fossil and nuclear resources and to the yearly potential of renewable alternatives. The volume of each sphere represents the total amount of energy recoverable from the finite reserves and the energy recoverable per year from renewable sources. The amount of solar energy is many orders of magnitude greater than all the other sources combined. In the long term, solar can have an increasingly positive impact on our energy independence.

The combination of oil, natural gas, nuclear power, hydroelectric, coal, wood, and other alternative energies will end our dependence on foreign oil and establish a strong economic base. You can advocate "doubling-down" on using alternative energies, but without a strong economic base and stable economy from which to work, an increasing implementation and use will be a slow one at best. We should not overregulate our domestic energy production to the point at which we cannot accomplish anything. That situation will continue to perpetuate our lack of progress in controlling and developing U.S. energy independence. Only with a strong economy will the United States be able to apply more renewable energy sources.

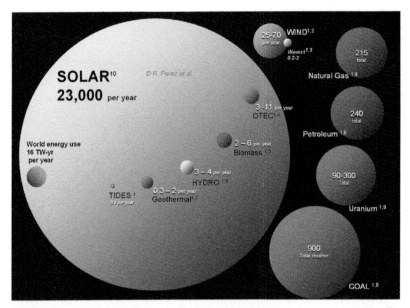

FIGURE 10.1 Comparing finite and renewable planetary energy reserves (Terawatt-years). Total recoverable reserves are shown for the finite resources. Yearly potential is shown for the renewables. Note: 1. Heckeroth, S. Renewables.com. Adapted from Christopher Swan, 1986. Sun Cell, Sierra Club Press. 2. Archer, C., Jacobson, M. Evaluation of Global Wind Power. Stanford University, Stanford, CA. 3. World Energy Council. 4. Nihous, G., December 2005. An order-of-magnitude estimate of ocean thermal energy conversion (OTEC) resources. Journal of Energy Resources Technology. 127 (4), 328–333. 5. Whittaker, R., 1975. The Biosphere and Man—In Primary Productivity of the Biosphere. Springer-Verlag, pp. 305–328. ISBN 0-3870-7083-4. 6. Environmental Resources Group, LLC, http://www.erg.com.np/hydropower_global.php. 7. MIT/INEL The Future of Geothermal Energy – Impact of Enhanced Geothermal Systems [EGS] on the U.S. in the 21st Century, http://www1.eere.energy.gov/geothermal/egs_technology.html, based on estimated energy recoverable economically in the next 50 years. Ultimate high-depth potential would be much higher. 8. BP Statistical Review of World Energy (2007). 9. Wise Uranium Project, http://www.wise-uranium.org/stk.html?src=stkd03e. 10. Price, R., Blaise, J.R., 2002. Nuclear fuel resources: Enough to last? NEA updates, NEA News 2002, No. 20.2. 11. Solar energy received by emerged continents only, assuming 65% losses by atmosphere and clouds. Source: With permission, courtesy of Perez & Perez. (For color version of this figure, the reader is referred to the online version of this book.)

Over the course of millions of years, the sun has provided us with stored chemical energy in the form of fossil fuels, which constantly are being depleted, and this depletion, as well as our dependence on foreign energy, is responsible for escalated social and economic costs. To curtail escalating fiscal adversities, the direct application of the sun's radiant energy to alternative conversion processes, such as PV, photochemical, thermionic, thermoelectric, and heat must be continuously developed and utilized. An economic first application for existing green-energy alternatives involves using solar collectors to convert the sun's radiant energy into heat and electrical energy.

Mathematical Tables, Relationships, and Conversions

MATHEMATICAL TABLES

Prefixes	atto	a	one–quintillionth	0.000 000 000 000 000 001	10^{-18}
	femto	f	one–quadrillionth	0.000 000 000 000 001	10^{-15}
	pico	p	one–trillionth	0.000 000 000 001	10^{-12}
	nano	n	one–billionth	0.000 000 001	10^{-9}
	micro		one–millionth	0.000 001	10^{-6}
	milli	m	one–thousandth	0.001	10^{-3}
	centi	c	one–hundredth	0.01	10^{-2}
	deci	d	one-tenth	0.1	10^{-1}
	uni		one	1.0	10^{0}
	deka	da	ten	10.0	10^{1}
	hecto	h	one hundred	100.0	10^{2}
	kilo	k	one thousand	1000.0	10^{3}
	mega	M	one million	1000 000.0	10^{6}
	giga	G	one billion	1000 000 000.0	10^{9}
	tera	T	one trillion	1000 000 000 000.0	10^{12}

Equivalent Measurements

Length	
1 in	25.4 mm
1 in	2.54 cm
1 mm	0.03937 in
1 mm	0.00328 ft
1 ft	304.8 mm

Continued...

Length	
1 m	3.281 ft
1 yard	0.914 m
1 km	0.621 mile
1 mile	1.609 km
1 μm	0.000001 m

Area	
1 in^2	6.4516 cm^2
1 ft^2	0.0929 m^2
1 mile^2	2.590 m^2
1 cm^2	0.155 in^2
1 m^2	10.764 ft^2
1 km^2	0.386 mile^2

Volume	
1 gallon (U.S.)	3.785 l
1 l	0.264 gallons (U.S.)
1 cm^3	0.061 in^3
1 barrel of oil	42 gallon (U.S.)
1 gallon of water at 60 °F	8.33 lb

Mass	
1 kg	1000 g = 2.205 lb
1 metric ton = 1000 kg	0.984 ton
1 in^3 of water (60 °F)	0.07355 in^3 of mercury (32 °F)
1 in^3 of mercury (32 °F)	0.4905 lb
1 in^3 of mercury (32 °F)	13.596 in^3 of water (60 °F)

Density	
1 lb/ft^3	16.0184 kg/m^3
1 kg/m^3	0.06243 lb/ft^3

Pressure	
1 std atm	14.696 lb/in^2 760 mmHg
1 bar	0.987 atm
1 lb/in^2 (psi)	2.307 ft of water 0.06805 atm

Energy	
1 BTU	1.055×10^3 J 778.169 ft-lb 252 cal
1 Calorie = 4.187 J	0.003968 BTU
1 W = 1 J/s	0.00134 hp
1 kWh	3.6×10^6 J
1 hp	550 ft-lb/s 2546.4 BTU/hr
1 Ly/hr = 11.63 W/m²	3.687 BTU/ft²

Fuel to Energy	
1 kWh	3413 BTU
1 gallon of oil	138,000 BTU
1 Therm	100,000 BTU 100 ft³ natural gas
1 gallon LP gas	93,000 BTU
1 cord mixed hardwood	24 MBTU
1 cord mixed softwood	15 MBTU
1 lb of coal	12,500 BTU

Physical Constants	
Acceleration of gravity (g)	32.174 ft/s² 980.665 cm/s²
Pi (π)	3.1415926536
Base of natural logarithms (e)	2.7182818285
Absolute zero = −273.15 °C	−459.67 °F
°C	5/9 (F − 32)
°F	9/5 C + 32

MATHEMATICAL RELATIONSHIPS

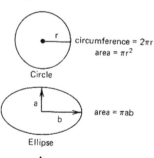

circumference = $2\pi r$
area = πr^2

Circle

area = πab

Ellipse

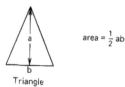

area = $\frac{1}{2}ab$

Triangle

area = a^2

Square

area = lw

Rectangle

Areas of common plane figures.

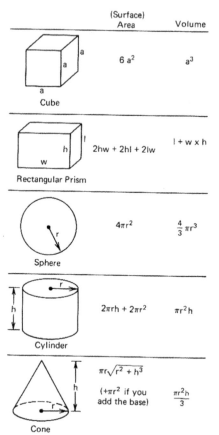

	(Surface) Area	Volume
Cube	$6\,a^2$	a^3
Rectangular Prism	$2hw + 2hl + 2lw$	$l + w \times h$
Sphere	$4\pi r^2$	$\frac{4}{3}\pi r^3$
Cylinder	$2\pi rh + 2\pi r^2$	$\pi r^2 h$
Cone	$\pi r\sqrt{r^2 + h^3}$ (+πr^2 if you add the base)	$\dfrac{\pi r^2 h}{3}$

Areas and volumes of common shapes.

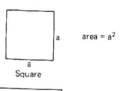

$\sin \theta = \dfrac{y}{r}$

$\cos \theta = \dfrac{x}{r}$

$\tan \theta = \dfrac{y}{x}$

Law of Exponents

$a^x \times a^x = a^{x+y}$ $\dfrac{1}{a^x} = a^{-x}$

$(ab)^x = a^x \times b^x$ $\dfrac{a^x}{a^y} = a^{x-y}$

$(a^x)^y = a^{xy}$ $a^0 = 1$

Laws of Logarithms
$\mathrm{Ln}(y^x) = x\,\mathrm{Ln}\,y$
$\mathrm{Ln}(ab) = \mathrm{Ln}\,a + \mathrm{Ln}\,b$
$\mathrm{Ln}\left(\dfrac{a}{b}\right) = \mathrm{Ln}\,a - \mathrm{Ln}\,b$

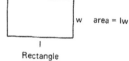

$\sin^2 \theta + \cos^2 \theta = 1$
$e^{i\theta} = \cos \theta + i \sin \theta$
$i = \sqrt{-1}$

Law of Cosines
$a^2 + b^2 - 2ab \cos \theta = c^2$

*Fundamental Identities and Reciprocal Relations
(A represents any angle)*

$\sin A = \dfrac{1}{\csc A}$, $\cos A = \dfrac{1}{\sec A}$, $\tan A = \dfrac{1}{\cot A}$

$\csc A = \dfrac{1}{\sin A}$, $\sec A = \dfrac{1}{\cos A}$, $\cot A = \dfrac{1}{\tan A}$

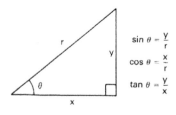

Trigonometric Relationships

MATHEMATICAL CONVERSION (ALPHABETICAL ORDER)

Multiply	By	To Obtain
Atmospheres	76.0	Centimeters-mercury
Atmosperes	29.92	Inches-mercury
Atmospheres	14.70	Pounds/inch^2
Atmospheres	1.058	Tons/feet^2
Barrels-oil	42	Gallons-oil
British thermal units	777.6	Foot-pounds
British thermal units	3.927×10^{-4}	Horsepower-hours
British thermal units	2.928×10^{-4}	Killowatt-hours
British thermal units	252	Calories
BTU/ft^2	0.271	Langleys
BTU/ft^2-hr	3.154	watts/meter^2
BTU/ft^2-hr	0.075	Calories/centimeter-second
BTU/ft^2-hr	0.00452	Langleys/minute
BTU/ft^2-hr-°F	0.488	Calories/hour-centimeter^2-°C
BTU/hr	0.216	Foot-pounds
BTU/hr	0.00039	Horsepower
BTU/hr	0.29283	Watts
Calories	0.00397	BTU
Calories/cm-s	13.272	BTU/feet-hour^2
Calories/hr-cm^2-°C	2.048	BTU/feet-hour^2-°F
Centimeters	0.3937	Inches
Centimeters	0.01	Meters
Centimeters	10	Millimeters
Centimeters-mercury	0.0132	Atmospheres
Centimeters-mercury	0.4460	Feet-water (4°C)
Centimeters-mercury	136.0	Kilograms/meter^2
Centimeters-mercury	27.85	Pounds/feet^2
Centimeters-mercury	0.1934	Pounds/inch^2
Centimeters/second	0.0328	Feet/second
Centimeters/second	0.0224	Miles/hour
Cubic centimeters	3.531×10^{-5}	Cubic feet
Cubic centimeters	0.0610	Cubic inches
Cubic centimeters	1×10^{-6}	Cubic meters
Cubic centimeters	1.3079×10^{-6}	Cubic yards
Cubic centimeters	2.642×10^{-4}	Gallons
Cubic centimeters	0.0010	Liters

Continued...

Multiply	By	To Obtain
Cubic feet	1728	Cubic inches
Cubic feet	0.0283	Cubic meters
Cubic feet	7.4805	Gallons
Cubic feet	28.32	Liters
Cubic feet/min	0.1247	Gallons/second
Cubic feet/min	0.4719	Liters/second
Cubic feet/second	448.831	Gallons/minute
Cubic ft of water at 60 °F	62.37	Pounds
Cubic inches	16.39	Cubic centimeters
Cubic inches	0.0005787	Cubic feet
Cubic inches	1.6387×10^{-5}	Cubic meters
Cubic inches	2.1433×10^{-5}	Cubic years
Cubic inches	0.004329	Gallons
Cubic inches	0.0164	Liters
Cubic meters	1×10^6	Cubic centimeters
Cubic meters	35.31	Cubic feet
Cubic meters	61,023	Cubic inches
Cubic meters	1.308	Cubic yards
Cubic meters	264.2	Gallons
Cubic meters	1000	Liters
Cubic yards	27	Cubic feet
Cubic yards	46.656	Cubic inches
Cubic yards	0.7645	Cubic meters
Cubic yards	202.0	Gallons
Cubic yards	764.5	Liters
Degrees (angle)	60	Minutes
Degrees (angle)	0.0174	Radians
Degrees (angle)	3600	Seconds
Degree/second	0.1667	Revolutions/minute
Degree/second	0.0028	Revolutions/second
Feet	30.48	Centimeters
Feet	12	Inches
Feet	0.3048	Meters
Feet	0.3333	Yards
Feet-water (4 °C)	0.8826	Inches-mercury
Feet-water	0.434	Pounds/inch^2
Feet-water	62.43	Pounds/feet^2

Multiply	By	To Obtain
Feet/min	0.5080	Centimeters/second
Feet/min	0.0114	Miles/hour
Feet/second	30.48	Centimeters/second
Feet/second	0.6818	Miles/hour
Foot-pounds	0.0013	British Thermal Units
Foot-pounds/min	3.030×10^{-5}	Horsepower
Foot-pounds/min	2.2597×10^{-5}	kilowatts
Gallons	3785	Cubic centimeters
Gallons	0.1337	Cubic feet
Gallons	231	Cubic inches
Gallons	0.0038	Cubic meters
Gallons	3.785	Liters
Gallons, Imperial	1.2009	U.S. Gallons
Gallons, U.S.	0.8327	Imperial gallons
Gallons-(Water at 60 °F)	8.33	Pounds-water
Gallons/min	0.00223	Cubic feet/second
Grams	980.7	Dynes
Grams	0.0022	Pounds
Grams/cm^3	0.0361	pounds/inch^3
Horsepower	2544	BTU/hour
Horsepower	550	Foot-pounds/second
Horsepower	0.7457	Kilowatts
Horsepower-hours	0.7457	Kilowatt-hours
Inches	2.540	Centimeters
Inches-mercury	0.033	Atmospheres
Inches-mercury	0.491	Pounds/inch^2
Inches-mercury	70.73	Pounds/feet^2
Inches-water	0.0735	Inches-mercury
Kilograms	980,665	Dynes
Kilograms	2.205	Pounds
Kilometers	3281	Feet
Kilometers	1000	Meters
Kilometers	0.6214	Miles
Kilometers	1094	Yards
Kilometers/hour	54.68	Feet/min
Kilometers/hour	0.5396	Knots
Kilowatts	56.8	BTU/minute

Continued...

Multiply	By	To Obtain
Kilowatts	737.6	Foot-pounds/second
Kilowatts	1.341	Horsepower
Kilowatt-hour	3413	BTU's
Kilowatt-hour	2.655×10^6	Foot-pounds
Kilowatt-hour	1.341	Horsepower-hour
Langleys	3.687	BTU/feet^2
Langleys/min	221.2	BTU/feet^2-hour
Liters	0.0353	Cubic feet
Liters	61.02	Cubic inches
Liters	0.0010	Cubic meters
Liters	0.2642	Gallons
Meters	3.281	Feet
Meters	39.37	Inches
Meters	0.001	Kilometers
Meters	1.094	Yards
Meters/min	0.06	Kilometers/hour
Meters/min	0.0373	Mile/hour
Meters/second	196.8	Feet/minute
Meters/second	3.281	Feet/second
Meters/second	0.03728	Miles/minute
Microns	1×10^6	Meters
Miles	5280	Feet
Miles	1.609	Kilometers
Miles/hr	88	Feet/minute
Miles/hr	1.467	Feet/second
Miles/hr	1.609	Kilometers/hour
Miles/hr	0.8690	Knots
Miles/min	2682	Centimeters/second
Miles/min	88	Feet/second
Miles/min	1.609	Kilometers/hour
Milligrams	0.001	Grams
Milliliters	0.001	Liters
Millimeters	0.1	Centimeters
Millimeters	0.0394	Inches
Ounces	0.0625	Pounds
Pounds	16	Ounces
Pounds	454	Grams

Multiply	By	To Obtain
Pounds/water	0.1198	Gallons
Pounds/in^3	1728	Pounds/feet^3
Pounds/ft	1488	Kilograms/meter
Pounds/in	178.6	Grams/centimeter
Pounds/ft.^{2}	4.882	Kilograms/meter^2
Pounds/in^2	0.0680	Atmospheres
Pounds/in^2	2.036	Inches-mercury
Quadrants (angle)	1.571	Radians
Quarts (liq.)	57.75	Cubic inches
Radians	57.30	Degrees
Radians	3438	Minutes
Radians	0.637	Quadrants
Radians/second	9.549	Revolutions/minute
Revolutions/second	360	Degrees/second
Revolutions/second	6.283	Radians/second
Revolutions/second	60	Revolutions/minute
Seconds (angle)	4.8481×10^{-6}	Radians
Square centimeters	0.0011	Square feet
Square centimeters	0.1550	Square inches
Square centimeters	0.0001	Square meters
Square centimeters	100	Square millimeters
Square feet	929.0	Square centimeters
Square feet	144	Square inches
Square feet	0.0929	Square meters
Square feet	3.5870×10^{-8}	Square miles
Square feet	0.1111	Square yards
Square inches	6.452	Square centimeters
Square inches	0.0069	Square feet
Square kilometers	1.0764×10^7	Square feet
Square kilometers	1×10^6	Square meters
Square kilometers	0.3861	Square miles
Square kilometers	1.1960×10^6	Square yards
Square meters	10.76	Square feet
Square meters	1.1960	Square yards
Square miles	2.590	Square kilometers
Square miles	3.0976×10^6	Square yards
Square millimeters	0.01	Square centimeters

Continued...

Multiply	By	To Obtain
Square millimeters	0.0016	Square inches
Square yards	9	Square feet
Square yards	0.8361	Square meters
Square yards	3.2283×10^{-7}	Square miles
Tons (metric)	2205	Pounds
Tons (short)	2000	Pounds
Tons of air conditioning	12,000	BTU/hour
Watts	0.7377	Foot-pounds/second
Watts	0.0013	Horsepower
Watts	0.001	Kilowatts
Watt-hours	3.413	British thermal units
Watt-hours	2655	Foot-pounds
Watt-hours	0.00134	Horsepower-hours
Watt-hours	0.001	Kilowatt-hours
Watts/m^2	0.317	BTU/feet^2-hour
Yards	91.44	Centimeters
Yards	3	Feet
Yards	36	Inches
Yards	0.9144	Meters

Worksheets

CHAPTER 4

Table 4.4b Worksheet for Collector Sizing, Energy Consumption, and Solar Contribution

Worksheet for Collector Sizing, Energy Consumption, and Solar Contribution

Latitude _____
Collector Tilt Angle _____

Line	Evaluation Factors		Jan.	Febr.	Mar.	Apr.	May	June	Jul.	Aug.	Sept.	Oct.	Nov.	Dec.	Total Per Year
								Month							
A	Days in month		31	28	31	30	31	30	31	31	30	31	30	31	365
B	No. of people														
[1]C	Hot water consumed (gallons; W_c)	Daily													
		Monthly													
D	Storage temperature (T_s)														
E	Inlet water temperature (T_i)														
F	Avg. temp. increase D-E ($T_s - T_i$)														
G	BTU requirement (daily) 8.33* C*F														
															Yearly average
[2]H	Collector system efficiency (n_s)														
[3]I	Available solar radiation (BTU/ft²-day)														
J	Array size (ft²) G ÷ (H*I)														4
													Actual total effective collector area installed (ft²)		

*Multiplication.
÷Division.

[1]Reference '20 + 20 + 15 + 15 + . . .' supposition.
[2]Assume an average of 0.5.
[3]Reference Table 4.3
[4]Effective collector aperture area required.

CHAPTER 5

Table 5.3 Sizing Method for the Determination of the Number of PV Modules

Sequence	Method	Calculation (i.e. Your Town*)	Result
Step 1	Annual Kilowatt-hours from 12 months of utility bills	Annual Kwh = _____ Kwh	_____ Kwh / year (Annual Demand)
Step 2	Average Daily Kwh	$\frac{Kwh}{year}$ X $\frac{1\ year}{365\ days}$	_____ Kwh/ day (Daily Demand)
Step 3	Divide Daily Demand by Average Sun Hours (*Table 5-2, Section 5.4)	$\frac{Kwh}{day}$ ÷ * $\frac{Hours}{day}$ = $\frac{KW}{day}$	_____ KW (System Size solar output required to yield 100% of Daily Demand)
Step 4	Multiply Step 3 by **1.15 to account for DC to AC inverter power and wire run losses (efficiencies)	_____ KW X 1.15 = _____ KW	_____ KW (System size output including system energy losses)
Step 5	Divide Daily Supplied Solar Energy System Output (Step 4) by the CEC wattage rating output per solar module	(Selection of a solar module with a PTC/CEC wattage output of _____ watts where _____ KW = _____ watts) _____ Watts ÷ _____ $\frac{watts}{panel}$ = _____	Total Number of Photovoltaic Modules = _____ (Number of modules required to produce 100% of electrical demand)

* Sun hours per day; National Average for Billings, MT. (See Table 5-2, Section 5.4)
** Multiply the solar output by 1.15 to adjust for efficiency losses in order to determine the number of modules required to produce 100% of the energy demand. The inverse is true if the number of modules is known due to limited roof area, in which case the known output for the array would be multiplied by .85 in order to determine the actual output of the array assuming efficiency losses of 15%.

CHAPTER 7

Typical Pump Operational Costs (See Table 7.4, Section 7.4, as an Example)

Horse-Power of Pump or Circulator	Annual Operating Cost of Solar DHW at Various Electric Rates (3)		
	(1) **kW Required**	**(2)** **Annual kWh** **(Based upon 8 h/day)**	**(3)** **(Annual Operating Cost at _____ ¢ per kWh Rate)** **Note: Multiply Column (2) by Column (3) Rate**
1/25	0.085	248.2	
1/20	0.098	286.2	
1/12	0.185	540.2	
1/4	0.420	1226.4	
1/3	0.530	1547.6	

CHAPTER 7

Table 7.1b Worksheet to Determine Solar Energy Contribution to DHW Energy Requirements

Worksheet to Determine Solar Energy Contribution to DHW Energy Requirements

Latitude _____

Collector Tilt Angle _____

[1]Effective Collector Aperture Area _____ ft^2

Line	Evaluation Factors		Jan.	Feb.	Mar.	Apr.	May	Jun.	Jul.	Aug.	Sept.	Oct.	Nov.	Dec.	Total per Year
A	Days in month		31	28	31	30	31	30	31	31	30	31	30	31	365
B	BTU requirement	[1]Daily (BTU)													
		Monthly (MBTU)													
C	Available solar radiation (BTU/ft^2)	[1]Daily													
		Monthly													
D	Collector array output (MBTU's) (Collector area), *C														
E	Mean percentage of possible sun (Table 7.2)														
F	Collector array output E * D (MBTU)														
G	[1]Collector system efficiency, Ns														
H	Estimated system output, G * F (MBTU)														
I	Percentage (%) solar energy contribution, $H \div B$														

[1]From Table 4.4b

*Multiplication.

÷ Division.

CHAPTER 7

(From Table 7.5 Energy Costs & Savings Realized (System Output of _____ MBTUs at $_____ per kWh)

| Years | Conventional Electrical DHW Fuel Costs at 5% per Year Inflation | | Solar Operational and Maintenance Costs at 5% per Year Inflation | | Savings Realized From Solar versus Conventional Electric | |
	(1) Yearly	(2) Cumulative	(3) Yearly	(4) Cumulative	(1) − (3) Yearly	(2) − (4) Cumulative
1						
2						
3						
4						
5						
6						
7						
8						
9						
10						
11						
12						
13						
14						
15						
16						
17						
18						
19						
20						
	Total		Total		Total	

Note: Reference Table 7.4 and include Operational and Maintenance Costs.

CHAPTER 8

(Developed from Table 8.3) Energy Costs at ____ cents/kWh (e.g. Table 8.2) for _____ kW and Savings Realized from electricity produced by a PV Array

Years	Conventional Electrical Demand Costs at 5% per Year Inflation		Solar PV Module Losses at 1.0% Output Degradation per Year		Savings Realized from Solar PV versus Electric Utility	
	(1) Yearly	(2) Cumulative	(3) Yearly	(4) Cumulative	(5) Yearly (1)–(3)	(6) Cumulative (2)–(4)
1						
2						
3						
4						
5						
6						
7						
8						
9						
10						
11						
12						
13						
14						
15						
16						
17						
18						
19						
20						
	Total		Total		Total	

Note: Reference Tables 8.2, Section 8.1 and 8.3, Section 8.3 as Examples and include Degradation Losses.

CHAPTER 9

(From Table 9.4) Visualizing the Capital Recovery Factor

Year	(1) Money Owed at Start of Year	(2) Interest Owed at End of Year at ___% (Insert Borrowed Interest Rate)	(3) (1) + (2) Principal and Interest Owed at End of Year	(4) Series (R) of Repayments	(5) (3) − (4) Money Owed at End of Year after Repayment	(6) (4) − (2) Recovery Capital
1						
2						
3						
4						
5						
6						
7						
8						
9						
10						
				Total		

Appendix C

The contents of this appendix have been excerpted from a report written by Dr Alan Carlin with permission. Only the executive summary, annotated as Section III, and a few detailed individual sections from that report are included. Each individual section of this appendix is entirely in Dr Carlin's words, and the organization of those sections has been arranged for the reader's convenience. References to particular figures and illustrations have been removed but are available within the author's original report. Abbreviations and references are available at the end of this appendix, and the entire comprehensive report is available at http://www.carlineconomics.com/archives/1.

COMMENTS ON DRAFT TECHNICAL SUPPORT DOCUMENT FOR ENDANGERMENT ANALYSIS FOR GREENHOUSE GAS EMISSIONS UNDER THE CLEAN AIR ACT

By Dr. Alan Carlin

IMPORTANT NOTE ON THE ORIGINS OF THESE COMMENTS

These comments were prepared during the week of March 9–16, 2009, and are based on the March 9 version of the draft Environmental Protection Agency (EPA) Technical Support Document (TSD) for the endangerment analysis for Greenhouse Gases under the Clean Air Act. On March 17, the Director of the National Center for Environmental Economics (NCEE) in the EPA Office of Policy, Economics, and Innovation communicated his decision not to forward these comments along the chain-of-command that would have resulted in their transmission to the Office of Air and Radiation, the authors of the draft TSD.

These comments (dated March 16) represent the last version prepared prior to the close of the internal EPA comment period as modified on June 27 to correct some of the nonsubstantive problems that could not be corrected at the time. No substantive change has been made from the version actually submitted on March 16.

It is very important that readers of these comments understand that these comments were prepared under severe time constraints. The actual time available was approximately 4–5 working days. It was therefore impossible to observe normal scholarly standards or even to carefully proofread the comments. As a result there are undoubtedly numerous unresolved inconsistencies and other problems that would normally have been resolved with more normal deadlines. No effort has been made to resolve any possible substantive issues; only a few of the more evident nonsubstantive ones have been resolved in this version. It should be noted, of course, that these comments represent the views of the author and not those of the U.S. Environmental Protection Agency or the NCEE.

I WHAT IS SCIENCE?

The first question is what science is. Science as used in these comments is the process of generating hypotheses and experimentally determining their validity by comparison with real world data—in other words, the application of the scientific method. I do not believe that science is writing a description of the world or the opinions of world authorities on a particular subject, or the number of scientists who agree on a particular issue. Science, I believe, is also not a statement of belief by scientific organizations. The question in my view is not what someone or some group believes but how what they believe corresponds to real world data. It is important to note that science evolves over time as new discoveries are made and new hypotheses are formulated and discarded. There is no such thing as permanent or settled science. Only continuing research can insure that important relationships are taken into account.

Richard Feynman (1965) expressed this as follows:

In general, we look for a new law by the following process. First, we guess it. Then we compute the consequences of the guess to see what would be implied if this law that we guessed is right. Then we compare the result of the computation to nature, with experiment or experience; compare it directly with observation to see if it works. If it disagrees with experiment it is wrong. It's that simple statement that is the key to science. It does not make any difference how beautiful your guess is. It does not make any difference how smart you are, who made the guess, or what his name is—if it disagrees with experiment (observation) it is wrong.

Fundamental to the science of global warming and of climate change is what determines the evident changes in global temperatures over time. Until this is firmly understood any attempt to determine the effects of particular changes in the climate environment such as increases in ambient greenhouse gas (GHG) levels on temperatures or human health and welfare is extremely risky since it runs the risk of being incorrect, with the result that any alleged endangerment may prove to be incorrect along with any actions that may be taken under the Clean Air Act as well.

II WHAT DETERMINES CHANGES IN GLOBAL TEMPERATURES?

Global temperatures have long fluctuated both in the short and long term. Until we clearly understand these fluctuations it is not possible to make any meaningful conclusions as to the cause of global warming. Numerous hypotheses have been offered, but they all cannot be correct since they differ greatly. One clue may be that there appears to be considerable cyclicality in temperatures over time; here is a brief synopsis of some of what I believe is known in terms of the length of the cycles involved:

Over 150 million year periods: There appears to have been a distinct approximately 150 million year cycle in Earth's temperatures. One explanation that has been offered is the change in level of galactic cosmic rays resulting from the Solar System's movements above and below the galactic plain resulting in higher cosmic ray levels when it is not in the plain.

Over 100,000 year periods: for the last 3 million years or so the Earth has gone through a succession of ice ages interspersed with relatively brief interglacial periods such as the one we are now in (called the Holocene). In the early part of this period they averaged about 40,000 years each but more recently they have averaged about 100,000 years in length. Global temperatures are believed to have been 5 to 10 °C less during ice ages than during interglacial periods. Various hypotheses have been proposed to explain this but the predominant view appears to be that it is due to changes in the Earth's orbit which change the intensity of the sun's radiation reaching the Earth (the so-called Milankovitch cycles). One problem with this explanation is that it does not explain the shift from 40,000 years to 100,000 year cycles. What appears evident, however, is that Earth's climate is unstable on the downside during the interglacial periods and unstable on the upside during ice ages. There appears to be something which has prevented the Earth from getting even colder than it has during ice ages or warming more than it has during interglacial periods. It is far from clear what these somethings are, but this asymmetry appears to have existed for at least 3 million years.

Over 1500 year (or so) periods: the Earth has had repeated cooler and warmer periods during the current interglacial (Holocene) period.

III EXECUTIVE SUMMARY

These comments are based on the draft TSD for Endangerment Analysis for Greenhouse Gas Emissions under the Clean Air Act (hereafter draft TSD) issued by the Climate Change Division of the Office of Atmospheric Programs on March 9, 2009. Unfortunately, because I was only given a few days to review this lengthy document these comments are of necessity much less comprehensive and polished than they would have been if more time had been allowed. I am prepared, however, to provide added information, more detailed comments on specific points raised, and any assistance in making changes if requested by Office of Air and Radiation (OAR).

The principal comments are as follows:

As of the best information I currently have, the GHG/CO$_2$ hypothesis as to the cause of global warming, which this draft TSD supports, is currently an invalid hypothesis from a scientific viewpoint because it fails a number of critical comparisons with available observable data. Any one of these failings should be enough to invalidate the hypothesis; the breadth of these failings leaves no other possible conclusion based on current data. As Feynman (1965) has said failure to conform to real-world data makes it necessary from a scientific view-point to revise the hypothesis or abandon it. Unfortunately this has not happened in the global warming debate, but needs to if an accurate finding concerning endangerment is to be made. The failings are listed below in decreasing order of importance in my view:

1. Lack of observed upper tropospheric heating in the tropics.
2. Lack of observed constant humidity levels, a very important assumption of all the United Nations, Intergovernmental Panel on Climate Change (IPCC) models, as CO_2 levels have risen.
3. The most reliable sets of global temperature data we have, using satellite microwave sounding units, show no appreciable temperature increases during the critical period 1978–1997, just when the surface station data show a pronounced rise. Satellite data after 1998 is also inconsistent with the GHG/CO_2/anthropogenic global warning (AGW) hypothesis.
4. The models used by the IPCC do not take into account or show the most important ocean oscillations which clearly do affect global temperatures, namely, the Pacific Decadal Oscillation (PDO), the Atlantic Multidecadal Oscillation (AMO), and the El Nino-Southern Oscillation (ENSO). Leaving out any major potential causes for global warming from the analysis results in the likely misattribution of the effects of these oscillations to the GHGs/CO_2 and hence is likely to overstate their importance as a cause for climate change.
5. The models and the IPCC ignored the possibility of indirect solar variability, which if important would again be likely to have the effect of overstating the importance of GHGs/CO_2.
6. The models and the IPCC ignored the possibility that there may be other significant natural effects on global temperatures that we do not yet understand. This possibility invalidates their statements that one must assume anthropogenic sources in order to duplicate the temperature record. The 1998 spike in global temperatures is very difficult to explain in any other way.
7. Surface global temperature data may have been hopelessly corrupted by the urban heat island effect and other problems which may explain some portion of the warming that would otherwise be attributed to GHGs/CO_2. In fact, the draft TSD refers almost exclusively to surface rather than satellite data.

The current draft TSD is based largely on the IPCC *AR4* report, which is at best 3 years out of date in a rapidly changing field. There have been important developments in areas that deserve careful attention in this draft. The list includes the following:

- Global temperatures have declined—extending the current downtrend to 11 years with a particularly rapid decline in 1907–1908; in addition, the PDO went negative in September 2007 and the AMO in January 2009, respectively. At the same time atmospheric CO_2 levels have continued to increase and CO_2 emissions have accelerated.
- The consensus on past, present and future Atlantic hurricane behavior has changed. Initially, it tilted towards the idea that AGW is leading to (and will

lead to) more frequent and intense storms. Now the consensus is much more neutral, arguing that future Atlantic tropical cyclones will be little different that those of the past.

- The idea that warming temperatures will cause Greenland to rapidly shed its ice has been greatly diminished by new results indicating little evidence for the operation of such processes.
- One of the worst economic recessions since World War II has greatly decreased GHG emissions compared to the assumptions made by the IPCC. To the extent that ambient GHG levels are relevant for future global temperatures, these emissions reductions should greatly influence the adverse effects of these emissions on public health and welfare. The current draft TSD does not reflect the changes that have already occurred nor those that are likely to occur in the future as a result of the recession. In fact, the topic is not even discussed to my knowledge.
- A new 2009 paper finds that the crucial assumption in the general circulation model used by the IPCC concerning strongly positive feedback from water vapor is not supported by empirical evidence and that the feedback is actually negative.
- A new 2009 paper by Scafetta and Wilson suggests that the IPCC used faulty solar data in dismissing the direct effect of solar variability on global temperatures. Other research by Scafetta and others suggests that solar variability could account for up to 68% of the increase in Earth's global temperatures.

These six developments alone should greatly influence any assessment of "vulnerability, risk, and impacts" of climate change within the United States, but are not discussed in the draft TSD to my knowledge. But these are just a few of the new developments since 2006. Therefore, the extensive portions of the EPA's Endangerment TSD which are based upon science from the IPPC *AR4* report are no longer appropriate and need to be revised before a TSD is issued for comments.

Not only is some of the science of the TSD out-of-date but there needs to be an explicit, in-depth analysis of the likely causes of global warming in my view. Despite the complexity of the climate system the following conclusions in this regard appear to be well supported by the available data:

1. By far the best single explanation for global temperature fluctuations appears to be variations in the PDO/AMO/ENSO. ENSO appears to operate in a 3–5 year cycle. PDO/AMO appear to operate in about a 60-year cycle. This is not really explained in the draft TSD but needs to be, or, at the very least, there needs to be an explanation as to why OAR believes that these evident cycles do not exist or why they are so unimportant as not to receive in-depth analysis.
2. There appears to be a strong association between solar sunspots/irradiance and global temperature fluctuations. It is unclear exactly how this operates, but it may be through indirect solar variability on cloud formation. This topic is not really explored in the draft TSD but needs to be since otherwise the effects of solar variations may be misattributed to the effects of changes in GHG levels.

3. Changes in GHG concentrations appear to have so little effect that it is difficult to find any effect in the satellite temperature record, which started in 1978.

4. The surface measurements (such as HadCRUT) are more ambiguous than the satellite measurements in that the increasing temperatures shown since the mid-1970s could either be due to the rapid growth of urbanization and the heat island effect or by the increase in GHG levels. However, since no such increase is shown in the satellite record it appears more likely that urbanization and the urban heat island effect and/or other measurement problems are the most likely cause. If so, the increases may have little to do with GHGs and everything to do with the rapid urbanization during the period. Given the discrepancy between surface temperature records in the 1940–1975 and 1998–2008 and the increases in GHG levels during these periods it appears even more unlikely that GHGs have as much of an effect on measured surface temperatures as claimed. These points need to be very carefully and fully discussed in the draft TSD if it is be scientifically credible.

5. Hence it is not reasonable to conclude that there is any endangerment from changes in GHG levels based on the satellite record, since almost all the fluctuations appear to be due to natural causes and not human-caused pollution as defined by the Clean Air Act. The surface record is more equivocal but needs to be carefully discussed, which would require substantial revision of the draft TSD.

6. There is a significant possibility that there are some other natural causes of global temperature fluctuations that we do not yet really understand and which may account for the very noticeable 1998 temperature peak which appears on both the satellite and surface temperature records. This possibility needs to be fully explained and discussed in the draft TSD. Until and unless these and many other inconsistencies referenced in these comments are adequately explained it would appear premature to attribute all or even most of what warming has occurred to changes in GHG/CO_2 atmospheric levels.

These inconsistencies between the TSD analysis and scientific observations are so important and sufficiently abstruse that in my view EPA needs to make an independent analysis of the science of global warming rather than adopting the conclusions of the IPCC and Climate Change Science Program (CCSP) without much more careful and independent EPA staff review than is evidenced by the draft TSD. Adopting the scientific conclusions of an outside group such as the IPCC or CCSP without thorough review by EPA is not in the EPA tradition anyway, and there seems to be little reason to change the tradition in this case. If their conclusions should be incorrect and EPA acts on them, it is EPA that will be blamed for inadequate research and understanding and reaching a possibly inaccurate determination of endangerment. Given the downward trend in temperatures since 1998 (which some think will continue until about 2030 given the 60-year cycle described), there is no particular reason to rush into decisions based on a scientific hypothesis that does not appear to explain much of the available data.

Finally, there is an obvious logical problem posed by steadily increasing U.S. health and welfare measures and the alleged endangerment of health and welfare discussed in this draft TSD during a period of rapid rise in at least CO_2 ambient levels. This discontinuity either needs to be carefully explained in the draft TSD or the conclusions changed.

IV SOLAR VARIABILITY

Prior to the advent of the IPCC and interest in the effects of increasing CO_2, the predominant view appears to have been that variations in global temperatures over periods less than 100,000 years were primarily due to solar variability since the Sun is Earth's major source of heat and light. A number of researchers have studied this over the years, and they have found some apparent relationships between sunspot cycles and global temperatures. Some (prominently Svensmark, 1998) have even developed a hypothesis to explain this apparent relationship. This hypothesis is roughly as follows:

Solar variability has been studied for at least 400 years. The general conclusion prior to 1990 was that the Sun is the major driver but there was little agreement as to the exact mechanism. But starting in 1990, the IPCC instead attributed warming to GHGs/humans. In 1998, however, Svensmark suggested a mechanism for indirect solar variability effects. Now many or even most global warming skeptics cite solar variability as the major cause and basis for their skepticism. In recent years there has been a furious debate/war on this issue. There has been some new research in recent years, however, some of which will be summarized in the following sections.

Predominant views prior to 1990:

- "Earth's temperature often seems to correlate directly with solar activity: when this activity is high the Earth is warm."
- "During the famous 'Little Ice Age' during the 17th century, the climate was notably cooler. This correlated with the Maunder Minimum on the sun, an interval of few sunspots and aurorae."
- "In the 11th and 12th centuries, a 'Medieval Maximum' in solar activity corresponded to the 'Medieval Optimum' in climate."
- "The 20th century has been marked by generally increasing levels of solar activity" (Hoyt and Schatten, 1997).

Indirect solar variability may be major/better explanation than GHGs. Although total solar irradiance (TSI) may not vary much, that does not rule out indirect effects of solar variability as the major cause of global climate changes. The impact of changes in solar eruptions, wind, and magnetic field may explain some or all known global climate changes during the Holocene together with volcanic eruptions. TSI even varies with sunspot cycles. Other researchers agree that solar variability may be related to temperature variations prior to mid-20th century. Svensmark (1998) hypothesized that the Sun's magnetic field varies with sunspots and determines the number of cosmic rays available to stimulate low level clouds on Earth.

In 2007, Jasper Kirkby of the European Organization for Nuclear Research published a review article which reached the following major conclusions:

- "Over the last few years diverse reconstructions of past climate change have revealed clear associations with cosmic ray variations recorded in cosmogenic isotope archives, providing persuasive evidence for solar or cosmic ray forcing of the climate."
  - "The high correlation of the temperature variations in the $\Delta14\,^\circ C$ record suggests that solar/cosmic ray forcing was a major driver of climate" (over the last 2000 years).
- "Two different classes of microphysical mechanisms have been proposed to connect cosmic rays with clouds:
  - Production of cloud condensation nuclei
  - Global electrical circuit in the atmosphere and, in turn, on ice nucleation and other cloud microphysical processes."
- "Considerable progress on understanding ion-aerosol-cloud processes has been made in recent years, and the results are suggestive of a physically-plausible link between cosmic rays, clouds and climate."

Conclusions were based on a broad review of the evidence for galactic cosmic rays (GCR) impact on climate using a number of different time periods and lines of evidence. The important points would appear to be the following:

1. GCRs are strongly related to global temperatures.
2. Solar activity modulates GCRs reaching earth, with the modulation related to sunspot cycles.

V ANOTHER POSSIBLE INCONSISTENCY: DO CHANGES IN CO_2 CAUSE CHANGES IN TEMPERATURE?

The IPCC (2007) argues that it is changes in ambient CO_2 levels that have and will largely determine temperature changes. A number of skeptics dispute this. One of their arguments is that changes in temperature have preceded changes in CO_2 by hundreds of years rather than the other way around over the last quarter million years (see Gregory, 2008, citing Caillon et al., 2003; and Singer, 2008, citing Fischer, 1999). They argue that this is incompatible with changes in CO_2 levels having any effect on temperature. According to Gregory (2009), "Logic demands that cause must precede effect. Increases in air temperature drive increases in atmospheric CO_2 concentration, and not vice versa."

VI SOME MAJOR INCONSISTENCIES IN THE SCIENCE OF GLOBAL WARMING THAT AT LEAST NEED TO BE EXPLAINED

In addition to the more recent inconsistencies discussed, there are a number of others of somewhat longer standing that at least need to be discussed in the draft TSD in my view. They are so serious, however, that I believe that there is a need to change the conclusions of the draft TSD. For a more complete list of inconsistencies that others have found see Gregory (2009) and Singer (2008). Gregory's list has approximately 30 items, few of which are addressed in the draft TSD. Although these lists themselves have not been peer-reviewed, many of the references have been. All these inconsistencies are included in these comments by reference. This includes the important missing heating of the upper troposphere in the tropics, which is briefly mentioned in the draft TSD. These lists and the references they cite, unless carefully and successfully answered in the draft TSD, largely eliminate the GHG hypothesis as a serious contender for explaining a significant part of the global warming that has occurred. This leaves the most fundamental issue as to what does cause global temperature fluctuations. It is possible that a chaotic system such as climate varies with little rhyme or reason, of course, but curiously there appear to be a few regularities in the data. Failure to consider a number of other factors beyond those that the IPCC and the draft TSD consider makes the draft TSD one-sided and unscientific in its discussion since it appears to pre-suppose the answer and the answer does not explain the observed fluctuations in global temperatures. Until the causes are clearly understood most any control effort (except stratospheric geoengineering—see Carlin, 2007 and 2008) is likely doomed to failure. It is only by taking a new and fundamental look at this question that a meaningful understanding of the endangerment can be reached. Although the hour may be late, it is only by doing so that an accurate endangerment TSD can be prepared.

ABBREVIATIONS

AGW	Anthropogenic global warning
AMO	Atlantic multidecadal oscillation
CCSP	Climate change science program
CERN	European Organization for Nuclear Research
ENSO	El Nino-southern oscillation
EPA	Environmental protection agency
GCM	General circulation model
GCR	Galactic cosmic rays
GHG	Greenhouse gas
GW	Global warning
HadCRUT	Data set of monthly instrumental temperature records formed by combining sea surface temperature records compiled by the Hadley Centre of the U.K. Met Office and the land surface air temperature records compiled by the Climatic Research Unit of the University of Anglia
IPCC	United Nations, Intergovernmental Panel on Climate Change
NCEE	National Center for Environmental Economics
OAR	Office of Air and Radiation
PDO	Pacific decadal oscillation
TSD	Technical support document
TSI	Total solar irradiance
UHI	Urban heat island

REFERENCES

Caillon, Nicholas, Jeffrey P. Severinghaus, Jean Jouzel, Jean-Marc Barnola, Jiancheng Kang, and Volodya Y. Lipenkov, March 14, 2003. "Timing of Atmospheric CO2 and Antarctic Temperature Changes Across Termination III," Science 299, 1728–31.

Carlin, Alan, June 2007. Global climate change control: is there a better strategy than reducing greenhouse gas emissions. University of Pennsylvania Law Review 155 (6), 1401–1497. Available at http://www.pennumbra.com/issues/pdfs/155-6/Carlin.pdf.

Carlin, Alan, Spring 2008. Why a different approach is required if global climate change is to be controlled efficiently or even at all. Environmental Law and Policy Review 32 (3), 685–757. Available at http://scholarship.law.wm.edu/cgi/viewcontent.cgi?article=1050& context=wmelpr.

Feynman, Richard, 1965. The Character of Natural Law. MIT Press. p. 156.

Fischer, H., et al., 1999. Carbon dioxide in the Vostok ice core. Science 283, 1712–1714.

Gray, William M., March 2009. Climate Change: Driven by the Ocean Not Human Activity. http://tropical.atmos.colostate.edu/Includes/Documents/Publications/gray2009.pdf.

Gregory, Ken, 2008. The Saturated Greenhouse Effect. Available at http://www.friendsofscien ce.org/assets/documents/The_Saturated_Greenhouse_Effect.htm.

Gregory, Ken, 2009. Climate Change Science. Available at http://www.friendsofscience.org/a ssets/documents/FOS%20Essay/Climate_Change_Science.html.

Hoyt, Douglas V., Kenneth, H. Schatten, 1997. The Role of the Sun in Climate Change. Oxford University Press.

Kirkby, Jasper, 2007. Cosmic rays and climate. Surveys in Geophysics 28, 333–375. Available as CERN-PH-EP/2008-005 dated March 26, 2008 at http://arxiv.org/pdf/ 0804.1938v1.pdf.

Scafetta, Nicola, Wilson, Richard C., March 3, 2009. ACRIM-gap and TSI trend issue resolved using a surface magnetic flux TSI model. Geophysical Research Letters 36, 1–5. L05701.

Singer, S. Fred (Ed.), 2008. Nature, Not Human Activity, Rules the Climate. Heartland Institute. Available at. http://heartland.org/policy-documents/nature-not-human-activity-rules-climate-pdf.

Svensmark, Henrik, 1998. Influence of cosmic rays on earth's climate. Physical Review Letters 81 (22), 5027–5030. Available at http://hep.physics.indiana.edu/~rickv/quarknet/ article2.pdf.

United Nations, Intergovernmental Panel on Climate Change, 2007. Fourth Assessment Report. AR4. Cambridge University Press. Available at http://www.ipcc.ch.

Notes

a. Uniform Resource Locators (URL), also known as web addresses, can change over periods of time. References in this appendix were updated June 2013. Should any of the URLs change over time, a global search by author and document name should locate the referenced material.

b. A Fifth Assessment Report (AR5) subsequent to reference 2 should be available for review in 2014.

c. Most of the comments in this Appendix are relative to the IPCC Report represented by Working Group I—Physical Science Basis.

Glossary

Absorber The surface of a collector (normally black) that absorbs the solar radiation and converts it to heat energy.

Absorptance (α) The ratio of solar energy absorbed by a surface to the solar energy striking it. Energy not absorbed is transmitted or reflected.

Acceleration The time rate of change of velocity.

Active solar system A solar heating or cooling system that requires external mechanical power to move the collected heat.

AFU Annual fuel utilization efficiency is a measure of how efficient a furnace is as a ratio of heat output compared to the total energy consumed.

AGW Anthropogenic global warning.

Altitude The angle of the sun's position in the sky with respect to the earth's horizontal.

Ambient temperature Temperature of the surroundings (i.e., for collectors, outdoor temperature).

Angle of incidence The angle measured between an incoming beam of radiation and a line drawn perpendicular to the surface that it strikes.

Anthropogenic Caused or produced by humans.

Aperture, solar The effective radiant energy collection area of a solar collector.

ASHRAE The American Society of Heating, Refrigerating, and Air Conditioning Engineers, Inc.

Azimuth The angular distance between true south and the point on the horizon directly below the sun.

Black body emitter An ideal body, which absorbs all radiation falling upon it and emits nothing.

British thermal unit (BTU) The quantity of heat energy needed to raise the temperature of one pound of water to $1\,°F$.

Calorie The quantity of heat energy needed to raise the temperature of $1\,g$ of water to $1\,°C$.

CEC Protocol for testing PTC testing standards adopted by the California Energy Commission. Reference PTC.

Centrifugal pump A type of pump, which has blades that rotate and whirl the fluid around so that it acquires sufficient momentum to discharge from the pump body.

Closed loop Any loop in a system that is not exposed to the atmosphere.

Coefficient A number that serves as a measure of some property or characteristic.

Collector Any of a wide variety of solar devices used to collect radiant energy and convert it to heat or electricity.

Collector efficiency The ratio of the heat energy or electricity extracted from a collector to the solar energy striking the cover, expressed in percent.

Collector tilt The angle between the horizontal plane and the solar collector plane.

Concentrating collector A device that concentrates the sun's rays on an absorber surface, which is significantly smaller than the overall collector area.

Conductance See Thermal conductance.

Conduction The transfer of heat energy through a material by the motion of adjacent atoms and molecules.

Conductivity The ease with which heat will flow through a material as determined by its physical characteristics.

Convection, forced Heat transfer through moving currents of air or liquid induced mechanically by a pump or blower in order to increase mass flow rates and velocities to yield a maximum heat transfer.

Convection, natural Heat transfer through moving currents of air or liquid as a result of thermal gradients and resulting density differences creating the necessary mass flow to promote heat transfer.

Corrosion Deterioration of metal by the chemical action of a fluid or components of a fluid.

CPI Consumer Price Index – An index of the change in the price of consumer goods and services from one base period to another.

Declination The angle between the plane of the earth's orbit and the equatorial plane.

Demand load Domestic water heating or electricity needs to be supplied by solar or conventional energy.

Density Weight per unit or mass per unit volume.

Design life The period of time for which a solar energy system is expected to perform its intended function without requiring major maintenance or replacement.

DHW Abbreviation for Domestic Hot Water. Also referred to as SHW (Solar Hot Water) and Solar Water Heating (SWH).

Diffuse radiation Solar radiation that has been scattered by clouds and particles in the atmosphere and casts no shadow. Flat plate collector can absorb it but concentrating collectors cannot.

Direct-beam radiation Solar energy received at the earth's surface uninterrupted by particles in the atmosphere and casting shadows on a clear day.

Ecliptic plane The plane of the earth's orbit extended to meet the celestial sphere.

Effectiveness See Heat exchanger effectiveness.

Efficiency The ratio of the useful energy supplied by a system (output) to the energy supplied to the system (input).

Elastomer Any of various elastic substances resembling rubber.

Electron An elementary particle consisting of a charge of negative electricity.

Emissivity (ε) A measure of the thermal energy reradiated from a solar collector surface as a fraction of the energy which would be radiated by a totally black body surface at the same temperature.

Energy Defined as the ability to do work. A conserved quantity, which is neither created nor destroyed. It can, however, be converted from one form to another or interconverted with matter according to Einstein's equation, $E=mc^2$ where m is mass and c is the speed of light.

Equinox The point of intersection of the ecliptic and celestial equator when the declination is zero.

External manifold A distribution pipe that runs outside the collector housing and connects to the inside header of each absorber plate.

Fill-drain assembly Comprised of two boiler drains and a check valve; installed for the filling and draining of transfer fluids in a closed loop freeze resistant system.

Flash point The temperature at which fluid vapors will flashover if an ignition source is present.

Flat plate collector A solar collector that converts sunlight to heat or electricity on a plane surface without the aid of reflecting surfaces to concentrate the sun's rays.

Fluid Any substance, gas, or liquid, used to capture heat in the collector and transport the energy from the point of collection to storage or direct use.

Flux Magnetic field lines.

Fossil fuels Combustible substances of organic origin established in past geologic ages consisting of hydrocarbons formed from the decay of vegetation under heat and pressure (coal, oil, natural gas).

Fusion The union of atomic nuclei to form heavier nuclei resulting in the release of enormous quantities of energy when certain light elements unite.

Galvanic corrosion Material degradation caused by an electrochemical reaction between two or more different metals in a system, which are not properly isolated from one another.

Galvanic series Metals ranked from electrically positive to electrically negative to provide a relative measure of "corrodability" of each metal when used in a multimetal system.

Gauge pressure That pressure measured above atmospheric pressure.

Generic Relating to or characteristic of an entire class.

GHG Greenhouse gas

Glazing A transparent/translucent sheet of glass/fiberglass that reduces heat loss from a solar collector and traps the thermal energy.

Greenhouse effect A retention of solar radiation in a solar panel whereby a cover glazing traps a layer of still air next to the absorber plate and reduces the convection heat loss. In terms of environmental considerations, it is a phenomenon whereby the earth's atmosphere traps the solar radiation by the presence of gases in the atmosphere such as carbon dioxide, water vapor, and methane, allowing incoming sunlight to pass through but absorbing heat radiated back from the earth's surface.

Grid An interconnected system for the distribution of electricity over a wide area through a network of high-tension cables and power stations.

Head For pumping considerations, the vertical rise to the highest point of a piping system.

Header The pipe that runs across the top or bottom of an absorber plate, gathering or distributing the heat transfer fluid to or from the risers that run across the absorber surface. (Also called manifolds if connected internal to each collector housing.)

Heat The sum total of all molecular energy of a body; a vector quantity.

Heat exchanger A device that transfers heat from one substance to another without mixing the two.

Heat exchanger effectiveness The ratio of the actual rate of heat transfer to the theoretical maximum rate of heat transfer.

Heat transfer medium Air or liquid that is heated and used to transmit energy from its point of collection to its point of storage and/or end use.

Hydrostatic Relating to liquids at rest or to the pressure they exert or transmit.

Inflation A persistent increase in the level of consumer prices or persistent decline in the purchasing power of money.

Infrared radiation Electromagnetic radiation from the sun that has wavelengths slightly longer than visible light, not visible to the naked eye.

Insolation The total amount of solar energy received at the earth's surface at any location and time (BTU/ft^2-h).

Insulation A material with a high thermal resistance (R) to heat flow.

Intercept The distance from the origin to a point where a graph crosses a coordinate axis.

Internal manifold A distribution pipe that connects the headers of the collectors internally and in turn becomes the header itself.

IPCC United Nations, Intergovernmental Panel on Climate Change.

Isogonic chart A chart depicting magnetic compass deviations from true south.

Index of refraction The measure of the bending of a ray of light when passing from one medium to another (ratio of the speed of light in a vacuum divided by the speed of light in a medium).

Inverter An electrical power converter that changes direct current to alternating current.

Kilowatt-hour (KWh) The amount of energy equivalent to1 kilowatt of power being used for 1 h (3413 BTU).

Kinetic energy Energy of motion.

Langley A unit of measure of insolation named for the American astronomer Samuel P. Langley ($3.687 BTU/ft^2$).

Latitude Referring to a point on the earth as determined by an angle formed by a line intersecting the center of the earth to a particular point on the its surface and the plane cutting the earth at the equator.

Liquid-type collector A collector that uses a liquid as the heat transfer fluid.

Longitude Referring to a point on the earth as determined by an angle formed by the intersection of a line from the center of the earth to a particular point on the earth's surface and the plane cutting vertically through the center of the earth.

Manifold See Header.

Mercator projection A graphical depiction of altitude and azimuth onto a flat map for each variation of latitude. Used for plotting obstacles, which might block energy collection in the "solar window".

Miscible Capable of being mixed.

NABCEP North American Board of Certified Energy Practitioners.

Net metering A renewable energy incentive allowing the energy produced from photovoltaic arrays to be fed into the utility power grid so that retail credit is received for a portion of the electricity generated.

Nonselective surface An absorber coating that absorbs most of the incident sunlight but which emits a high level of thermal radiation in return. Typically, it is a flat-black paint.

Normalized curve Conformed to a standardized reference base.

NREL National Renewable Energy Laboratory previously known as the Solar Energy Research Institute.

Opaque Impervious to forms of radiant energy other than visible light.

Open loop Any loop in a system that is vented to the atmosphere.

Ordinate The vertical coordinate of a point in a two-dimensional plane, parallel to the y-axis.

Orientation Number of degrees to the east or west of south that a solar collection surface faces.

Overall coefficient of transmittance (U-value) The reciprocal value of the sum of thermal resistances (BTU/hr-ft^2-°F). This value is the combined thermal conduction value of all the materials in a cross section including air spaces and air films. The lower the U-value, the higher the insulating value.

Passive solar system A system that uses gravity, heat flows, evaporation, or other naturally occurring phenomena without the use of external mechanical devices to transfer the collected energy (i.e., south facing windows).

Payback The time at which the initial cost and annual operating and maintenance expenses of a solar energy system equal the total savings generated by the system when compared with conventional energy sources. Both systems costs are computed at compounded interest rates of inflation for the same amount of energy generated.

Peak sun hours The equivalent number of hours per day when solar insolation averages 1 KW/m^2.

Percentage of possible sunlight The percentage of daytime hours during which there is enough direct solar radiation to cast a shadow.

pH Measure of solution acidity.

Photochemical The effect of radiant energy in producing chemical changes.

Photon A quantum of electromagnetic radiation that has zero rest mass and an energy of Plank's constant times the frequency of the radiation. Photons are generated in collisions between nuclei or electrons and in any other process in which an electrically charged particle changes its momentum. Photons can also be absorbed by any charged particle.

Photovoltaic Concerning the generation of an electromotive force when radiant energy falls on the boundary between dissimilar substances.

Photovoltaic cell An electrical device that converts light energy directly into electricity.

Pitch The ratio of vertical rise to horizontal span where rise is the distance from the attic floor to roof peak and span is the width of the house.

Plank's constant A universal constant of nature relating the energy of a quantum of radiation to the frequency of the oscillator from which it was emitted. Its numerical value is $6.6260755 \times 10^{-27}$ erg-seconds $= 6.6260755 \times 10^{-34}$ joules per hertz.

Potable water Water free from impurities present in amounts sufficient to cause disease or harmful physiological effects and conforming in its bacteriological and chemical quality to the requirements of the Public Health Service Drinking Water Standards or the regulations of the public health authority having jurisdiction.

Pound The basic unit of force in the English system of measure, defined as the force that gives a standard pound (0.4535924277 kg) an acceleration equal to the standard acceleration of earth's gravity, which is 32.174 ft/s^2.

Pressure The ratio of force per unit area.

Pressure drop The loss in static pressure through a component such as a heat exchanger, length of pipe, or duct, which may include fittings such as elbows or the combined losses throughout the entire length of fluid flow travel. Normally stated in terms of pounds per square inch (psi).

Proton An elementary particle consisting of a charge of positive electricity.

PTC Photovoltaic industry protocol called the "Photovoltaics for Utility Scale Applications" Test Conditions known as PVUSA Test Conditions or PTC used for testing comparisons of PV modules.

Radiant energy The flow of energy across open space via electromagnetic waves (i.e., visible light).

Reflectivity The ratio of the radiant energy reflected from a surface, to the radiant energy incident upon that surface.

Refraction The bending of light or sound waves when passing from one medium to another of different optical density.

Reradiation Radiation resulting from the emission of previously absorbed radiation.

Risers The flow channels or pipes that distribute the heat transfer liquid from the headers across the surface of an absorber plate.

R-value See Thermal resistance.

Selective surface A surface that absorbs radiation of one wavelength (i.e., visible light) but emits little radiation of another wavelength (i.e., infrared), thereby reducing heat loss.

Sky vault The entire projection of the sun's path at any particular latitude. (See Sun path diagram)

Slope The ratio of vertical rise to horizontal run where rise is the distance from the attic floor to roof peak and the run is the horizontal distance from the roof peak to the end of the roof section.

Solar constant The average amount of solar radiation reaching the earth's outer atmosphere ($436.5\,BTU/ft^2$-hr; $\pm3.5\%$)

Solar noon The instant of time the sun's position is true south (azimuth is $0°$) and the altitude is a maximum for the day.

Solar radiation (solar energy) Electromagnetic radiation emitted by the sun. The visible part of this spectrum ranges from long red to short violet wavelengths.

Solar window An outline of an area in the sky for a particular latitude through which a maximum amount of direct solar radiation reaches the collectors during any particular time of year and day.

Solstice The time at which the sun reaches its greatest declination, north or south.

SPCAF Single Payment Compound Amount Factor.

Specific heat The quantity of heat, in BTUs, needed to raise the temperature of one pound of material to $1\,°F$ (BTU/lb-$°F$).

Spectral distribution An energy curve or graph that shows the variation of radiant energy in relation to wavelengths.

SRCC Solar Rating and Certification Corporation—An independent third party certification organization that administers national rating programs for solar energy equipment providing a means to compare thermal performance of solar DHW collectors.

Stagnation A no-flow condition.

Standby heat loss Heat lost through storage tank and piping walls under no flow conditions.

STC Standard Test Conditions—A manufacturer's testing protocol whereas PTC or CEC protocols are preferred.

Sun path diagram (solar window) A circular projection of the sky vault, similar to a map, which can be used to determine solar position and to calculate shading.

Temperature An indicator of the intensity or degree of heat stored in a body; a scalar quantity.

Temperature gradient A change in temperature in a specific direction.

Thermal conductance (C) A property of a material equal to the quantity of heat per unit time that will pass through a unit area of the material when a unit average temperature is established between the surfaces (BTU/hr-ft^2-$°F$).

Thermal conductivity (K) A measure of the ability of a material to permit the flow of heat (BTU-in/hr-ft^2-$°F$).

Thermal inertia The tendency of a large mass to remain at the same temperature or to fluctuate only very slowly when acted upon by external sources.

Thermal resistance (R-value) A measure of the ability of a material to resist the flow of heat; the higher the R-value, the greater the insulating value of the material (hr-ft^2-$°F/BTU$).

Thermistor A sensing device that changes its electrical resistance with changes in temperature. Used with differential controllers and control monitors to supply collector and storage tank temperature information.

Thermoelectric Involving relations between temperature and the electrical condition in a metal or in contacting metals.

Thermionic Dealing with electrically charged particles emitted by an incandescent substance.

Thermosiphon The natural convection of heat through a fluid that occurs when a warm fluid rises and cool fluid sinks under the influence of gravity.

Tilt angle See Collector tilt.

Toxic fluids Gases or liquids that are poisonous, irritating, and/or suffocating, as classified in the Hazardous Substances Act, Code of Federal Regulations, Title 16, Part 1500.

Translucent Admitting and diffusing light so that objects beyond cannot be distinguished clearly.

Transmissivity (τ) The ratio of solar energy passed through a surface to the radiation striking it. Energy not transmitted is either absorbed and/or reflected.

Transparent Having the ability to transmit light without appreciable scattering so that objects beyond are entirely visible.

Tropic of Cancer The latitude denoting the most northerly position of the sun in which the declination angle is +23.5°.

Tropic of Capricorn The latitude denoting the most southerly position of the sun in which the declination angle is −23.5°.

Ultraviolet radiation Electromagnetic radiation with wavelengths shorter than visible light.

U-value See Overall coefficient of transmittance.

Viscosity The readiness with which a fluid flows when acted upon by an external force (g/cm-s).

Wavelength The distance between the start and finish of an energy pulse.

References

American Solar Energy Society (ASES), Masia, S., December 2013. Allies emerge for solar. Solar Today. and "Shout out for solar unity", Solar Citizen – January 2014.

Caillon, Nicholas, Jeffrey P. Severinghaus, Jean Jouzel, Jean-Marc Barnola, Jiancheng Kang, and Volodya Y. Lipenkov, March 14, 2003. "Timing of Atmospheric CO2 and Antarctic Temperature Changes Across Termination III," Science 299, 1728–31.

Carlin, A., April 1, 2011. A multidisciplinary, science-based approach to the economics of climate change. International Journal of Environmental Research and Public Health 8 (4), 985–1031.

Conklin, P., Alley, R., Broecker, W., Denton, G., 2011. The Fate of Greenland, Lessons from Abrupt Climate Change. The MIT Press.

EPA. United States Environmental Protection Agency. Office of Atmospheric Programs, April 2010. Methane and Nitrous Oxide Emissions from Natural Sources. EPA430-R-10-001.

Federal Reserve Bank of Cleveland. http://www.clevelandfed.org/research/data/US-inflation/cpi.cfm (accessed 06.10.13).

Federal Reserve Economic Data, June 10, 2013. Federal Reserve Bank of St. Louis. May, Travis. Research Division.

Fischer, S., Muller-Steinhagen, H., 2001. Collector Test Method under Quasi-Dynamic Conditions According to the European Standard EN 12975-2. ISES Solar World Congress.

Halliday, D., Resnick, R., Walker, J., 1997. Fundamentals of Physics. John Wiley & Sons, Inc., New York.

Ingard, U., Kraushaar, W.L., 1961. Introduction to Mechanics, Matter, and Waves. Addison-Wesley Publishing Co., Inc.

International Solar Energy Society (ISES), Fischer, S., Muller-Steinhagen, H., Perers, B., Bergquist, P., 2001. Collector Test Method under Quasi-Dynamic Conditions According to the European Standard EN 12975-2. ISES Solar World Congress.

McConnell, C., 1969. Economics: Principles, Problems, and Policies. McGraw-Hill Co., New York.

National Geophysical Data Center of the Oceanic and Atmospheric Administration. http://www.ngdc.noaa.gov.

National Imaging and Mapping Agency, renamed National Geospatial-Intelligence Agency (NGA), author unlisted, NIMA agency personnel. http://msi.nga.mil/MSISiteContent/StaticFiels/Files/mv-world.jpg.

National Oceanic and Atmospheric Administration (NOAA) and National Climatic Data Center (NCDC), U.S. Department of Commerce.

National Renewable Energy Laboratory for U.S. Department of Energy, August 2002. How to Size a Grid-Connected Solar Electric System. DOE/GO-102002-1607.

Plante, R., 1983. Solar Domestic Hot Water: A Practical Guide to Installation and Understanding. John Wiley & Sons, Inc., New York.

Shea, S.P., August 2012. Evaluation of Glare Potential for Photovoltaic Installations. Sunniva, Inc.

United Nations, Intergovernmental Panel on Climate Change, 2007. Fourth Assessment Report (AR4). Cambridge University Press.

U.S. Department of Agriculture, Forest Products Laboratory, State and Private Forestry Technology Marketing Unit, WOE-3, TechLine. 7/2004.

U.S. Department of Energy, September, 2010. Procuring Solar Energy: A Guide for Federal Decision Makers.

U.S. Department of Energy, April, 2013. Energy Efficiency & Renewable Energy, Energy Basics. Flat Plate Photovoltaic Systems.

U.S. Department of Labor, 2012. Consumer Price Index.

Washburn, J., May 2011. Microinverters are Launching a Solar Renewable Revolution. NASA Tech Briefs pp. 16–18.

Wiles, J., 1996. Photovoltaic Power Systems and the National Electrical Code: Suggested Practices. Sandia Report SAND96-2797.UC-1290. The Photovoltaic Systems Assistance Center-Sandia National Laboratories.

Wolfson, R., Pasachoff, J.M., 1999. Physics. Addison Wesley Longman, Inc.

DATA WEBSITES OF INTEREST

Database of State Incentives for Renewable Energy (DSIRE) for State Solar Tax Rebates. www.dsireusa.org/solar/.

Current average retail prices of electricity for each State available at U.S. Energy Information Administration (EIA). www.eia.gov/electricity.monthly.

PV module array power output estimator referred to as PV Watts™ Grid Data Calculator available at the Renewable Resource Energy Data Center at the National Renewable Energy Laboratory (NREL). www.nrel.gov/rredc.

On-line calculator for magnetic fields from the National Geophysical Data Center of the Oceanic and Atmospheric Administration. www.ngdc.noaa.gov/geomag.

Collector certification ratings available from the Solar Rating and Certification Corporation (SRCC). www.solar-rating.org.

Fuel Value and Power Calculators provides a tool that can be used to compare typical costs of various fuels; Available from the U.S. Department of Agriculture – Forest Products Laboratory; www.fpl.fs.fed.us/documnts/techline/fuel-value-calculator.pdf.

NOTE

Uniform Resource Locators (URL), also known as web addresses, can change over periods of time. References to Data Websites of Interest were updated June 2013. Should any of the URLs change over time, a global search by subject and document name should locate the referenced material.

Index

Note: Page numbers with "*f*" denote figures; "*t*" tables.

Printed and bound by CPI Group (UK) Ltd, Croydon, CR0 4YY

03/10/2024

01040423-0003